P1

THE FILTER BUBBLE

Eli Pariser is the board president and former executive director of the five–million–member organization MoveOn.org. A pioneer in online politics, he is a senior fellow at the Roosevelt Institute and cofounder of Avaaz.org, one of the world's largest citizen organizations. His op-eds have appeared in *The Washington Post*, the *Los Angeles Times*, and *The Wall Street Journal*. He grew up in Lincolnville, Maine.

Praise for *The Filter Bubble*

"Well-timed . . . a powerful indictment of the current system."
—*The Wall Street Journal*

"Eli Pariser is no enemy of the Internet. The thirty-year-old online organizer is the former executive director and now board president of the online liberal political group MoveOn.org. But while Pariser understands the influence of the Internet, he also knows the power of online search engines and social networks to control exactly how we get information—for good and for ill."
—*Time*

"Chilling." —*The New York Review of Books*

"Fascinating . . . a compelling deep-dive into the invisible algorithmic editing on the Web, a world where we're being shown more of what algorithms think we *want* to see and less of what we *should* see." —TheAtlantic.com

"'Personalization' sounds pretty benign, but Eli Pariser skillfully builds a case that its excess on the Internet will unleash an information calamity—unless we heed his warnings. Top-notch journalism and analysis."

—Steven Levy, author of *In the Plex: How Google Thinks, Works, and Shapes Our Lives*

"The Internet software that we use is getting smarter, and more tailored to our needs, all the time. The risk, Eli Pariser reveals, is that we increasingly won't see other perspectives. In *The Filter Bubble*, he shows us how the trend could reinforce partisan and narrow mindsets, and points the way to a greater online diversity of perspective."

—Craig Newmark, founder of craigslist

"Eli Pariser has written a must-read book about one of the central issues in contemporary culture: personalization."

—Caterina Fake, cofounder of Flickr and Hunch

"You spend half your life in Internet space, but trust me—you don't understand how it works. Eli Pariser's book is a masterpiece of both investigation and interpretation; he exposes the way we're sent down particular information tunnels, and he explains how we might once again find ourselves in a broad public square of ideas. This couldn't be a more interesting book; it casts an illuminating light on so many of our daily encounters."

—Bill McKibben, author of *Eaarth* and *The End of Nature*, and founder of 350.org

"*The Filter Bubble* shows how unintended consequences of well-meaning online designs can impose profound and sudden changes on politics. All agree that the Internet is a potent tool for change, but whether changes are for the better or worse is up to the people who create and use it. If you feel that the Web is your wide open window on the world, you need to read this book to understand what you aren't seeing."

—Jaron Lanier, author of *You Are Not a Gadget*

"For more than a decade, reflective souls have worried about the consequences of perfect personalization. Eli Pariser's is the most powerful and troubling critique yet."

—Lawrence Lessig, author of *Code v2, Free Culture,* and *Remix*

"Eli Pariser isn't just the smartest person I know thinking about the relationship of digital technology to participation in the democratic process—he is also the most experienced. *The Filter Bubble* reveals how the world we encounter is shaped by programs whose very purpose is to narrow what we see and increase the predictability of our responses. Anyone who cares about the future of human agency in a digital landscape should read this book—especially if it is not showing up in your recommended reads on Amazon."

—Douglas Rushkoff, author of *Life Inc.* and *Program or Be Programmed*

"In *The Filter Bubble*, Eli Pariser reveals the news slogan of the personalized Internet: Only the news that fits you we print."

—George Lakoff, author of *Don't Think of an Elephant!* and *The Political Mind*

THE FILTER BUBBLE

How the New Personalized Web
Is Changing What We Read
and How We Think

ELI PARISER

PENGUIN BOOKS

PENGUIN BOOKS

Published by the Penguin Group

Penguin Group (USA) Inc., 375 Hudson Street, New York,
New York 10014, U.S.A. • Penguin Group (Canada), 90 Eglinton
Avenue East, Suite 700, Toronto, Ontario, Canada M4P 2Y3 (a division
of Pearson Penguin Canada Inc.) • Penguin Books Ltd, 80 Strand, London
WC2R 0RL, England • Penguin Ireland, 25 St. Stephen's Green, Dublin 2, Ireland
(a division of Penguin Books Ltd) • Penguin Books Australia Ltd, 250 Camberwell
Road, Camberwell, Victoria 3124, Australia (a division of Pearson Australia Group
Pty Ltd) • Penguin Books India Pvt Ltd, 11 Community Centre, Panchsheel Park,
New Delhi – 110 017, India • Penguin Group (NZ), 67 Apollo Drive, Rosedale,
Auckland 0632, New Zealand (a division of Pearson New Zealand Ltd) •
Penguin Books (South Africa) (Pty) Ltd, 24 Sturdee Avenue,
Rosebank, Johannesburg 2196, South Africa

Penguin Books Ltd, Registered Offices: 80 Strand, London WC2R 0RL, England

First published in the United States of America by The Penguin Press,
a member of Penguin Group (USA) Inc. 2011
Published in Penguin Books 2012

3 5 7 9 10 8 6 4

ISBN 978-1-59420-300-8 (hc.)
ISBN 978-0-14-312123-7 (pbk.)

Printed in the United States of America
Designed by Chris Welch

To my grandfather, Ray Pariser, who taught me that scientific knowledge is best used in the pursuit of a better world. And to my community of family and friends, who fill my bubble with intelligence, humor, and love.

CONTENTS

INTRODUCTION

A squirrel dying in front of your house may be more relevant to your interests right now than people dying in Africa.

—*Mark Zuckerberg*, Facebook founder

We shape our tools, and thereafter our tools shape us.

—*Marshall McLuhan*, media theorist

Few people noticed the post that appeared on Google's corporate blog on December 4, 2009. It didn't beg for attention—no sweeping pronouncements, no Silicon Valley hype, just a few paragraphs of text sandwiched between a weekly roundup of top search terms and an update about Google's finance software.

Not everyone missed it. Search engine blogger Danny Sullivan pores over the items on Google's blog looking for clues about where the monolith is headed next, and to him, the post was a big deal. In fact, he wrote later that day, it was "the biggest change that has ever happened in search engines." For Danny, the headline said it all: "Personalized search for everyone."

Starting that morning, Google would use fifty-seven *signals*—everything from where you were logging in from to what browser you were using to what you had searched for before—to make guesses about who you were and what kinds of sites you'd like. Even if you were logged out, it would customize its results, showing you the pages it predicted you were most likely to click on.

Most of us assume that when we Google a term, we all see the same results—the ones that the company's famous Page Rank algorithm suggests are the most authoritative based on other pages' links. But since December 2009, this is no longer true. Now you get the result that Google's algorithm suggests is best for you in particular—and someone else may see something entirely different. In other words, there is no standard Google anymore.

It's not hard to see this difference in action. In the spring of 2010, while the remains of the Deepwater Horizon oil rig were spewing crude oil into the Gulf of Mexico, I asked two friends to search for the term "BP." They're pretty similar—educated white left-leaning women who live in the Northeast. But the results they saw were quite different. One of my friends saw investment information about BP. The other saw news. For one, the first page of results contained links about the oil spill; for the other, there was nothing about it except for a promotional ad from BP.

Even the number of results returned by Google differed—about 180 million results for one friend and 139 million for the other. If the results were that different for these two progressive East Coast women, imagine how different they would be for my friends and, say, an elderly Republican in Texas (or, for that matter, a businessman in Japan).

With Google personalized for everyone, the query "stem cells" might produce diametrically opposed results for scientists who support stem cell research and activists who oppose it. "Proof of climate change" might turn up different results for an environmental activist and an oil company executive. In polls, a huge majority of us assume search engines are unbiased. But that may be just because they're increasingly biased to share our own views. More and more, your computer monitor is a kind of one-way mirror, reflecting your own interests while algorithmic observers watch what you click.

Google's announcement marked the turning point of an important but nearly invisible revolution in how we consume information. You could say that on December 4, 2009, the era of personalization began.

WHEN I WAS growing up in rural Maine in the 1990s, a new *Wired* arrived at our farmhouse every month, full of stories about AOL and Apple and how hackers and technologists were changing the world. To my preteen self, it seemed clear that the Internet was going to democratize the world, connecting us with better information and the power to act on it. The California futurists and techno-optimists in those pages spoke with a clear-eyed certainty: an inevitable, irresistible revolution was just around the corner, one that would flatten society, unseat the elites, and usher in a kind of freewheeling global utopia.

During college, I taught myself HTML and some rudimentary pieces of the languages PHP and SQL. I dabbled in building Web sites for friends and college projects. And when an e-mail referring people to a Web site I had started went viral after 9/11, I was suddenly put in touch with half a million people from 192 countries.

To a twenty-year-old, it was an extraordinary experience—in a matter of days, I had ended up at the center of a small movement. It was also overwhelming. So I joined forces with a small civic-minded startup from Berkeley called MoveOn.org. The cofounders, Wes Boyd and Joan Blades, had built a software company that brought the world the Flying Toasters screen saver. Our lead programmer was a twenty-something libertarian named Patrick Kane; his consulting service, We Also Walk Dogs, was named after a sci-fi story. Carrie Olson, a veteran of the Flying Toaster days, managed operations. We all worked out of our homes.

The work itself was mostly unglamorous—formatting and sending out e-mails, building Web pages. But it was exciting because we were sure the Internet had the potential to usher in a new era of transparency. The prospect that leaders could directly communicate, for free, with constituents could change everything. And the Internet gave constituents new power to aggregate their efforts and make their voices heard. When we looked at Washington, we saw a system clogged with gatekeepers and bureaucrats; the Internet had the potential to wash all of that away.

When I joined MoveOn in 2001, we had about five hundred thousand U.S. members. Today, there are 5 million members—making it one of the largest advocacy groups in America, significantly larger than the NRA. Together, our members have given over $120 million in small donations to support causes we've identified together—health care for everyone, a green economy, and a flourishing democratic process, to name a few.

For a time, it seemed that the Internet was going to entirely redemocratize society. Bloggers and citizen journalists would single-handedly rebuild the public media. Politicians would be able to run only with a broad base of support from small, everyday donors. Local governments would become more transparent and accountable to their citizens. And yet the era of civic connection I dreamed about hasn't come. Democracy requires citizens to see things from one another's point of view, but instead we're more and more enclosed in our own bubbles. Democracy requires a reliance on shared facts; instead, we're being offered parallel but separate universes.

My sense of unease crystallized when I noticed that my conservative friends had disappeared from my Facebook page. Politically, I lean to the left, but I like to hear what conservatives are thinking, and I've gone out of my way to befriend a few and add them as Facebook connections. I wanted to see what links they'd post, read their comments, and learn a bit from them.

But their links never turned up in my Top News feed. Facebook was apparently doing the math and noticing that I was still clicking my progressive friends' links more than my conservative friends'—and links to the latest Lady Gaga videos more than either. So no conservative links for me.

I started doing some research, trying to understand how Facebook was deciding what to show me and what to hide. As it turned out, Facebook wasn't alone.

WITH LITTLE NOTICE or fanfare, the digital world is fundamentally changing. What was once an anonymous medium where anyone could be anyone—where, in the words of the famous *New Yorker* cartoon, nobody knows you're a dog—is now a tool for soliciting and analyzing our personal data. According to one *Wall Street Journal* study, the top fifty Internet sites, from CNN to Yahoo to MSN, install an average of 64 data-laden cookies and personal tracking beacons each. Search for a word like "depression" on Dictionary.com, and the site installs up to 223 tracking cookies and beacons on your computer so that other Web sites can target you with antidepressants. Share an article about cooking on ABC News, and you may be chased around the Web by ads for Teflon-coated pots. Open—even for an instant—a page listing signs that your spouse may be cheating and prepare to be haunted with DNA paternity-test ads. The new Internet doesn't just know you're a dog; it knows your breed and wants to sell you a bowl of premium kibble.

The race to know as much as possible about you has become the central battle of the era for Internet giants like Google, Facebook, Apple, and Microsoft. As Chris Palmer of the Electronic Frontier Foundation explained to me, "You're getting a free service, and the cost is information about you. And Google and Facebook translate that pretty directly into money." While

Gmail and Facebook may be helpful, free tools, they are also extremely effective and voracious extraction engines into which we pour the most intimate details of our lives. Your smooth new iPhone knows exactly where you go, whom you call, what you read; with its built-in microphone, gyroscope, and GPS, it can tell whether you're walking or in a car or at a party.

While Google has (so far) promised to keep your personal data to itself, other popular Web sites and apps—from the airfare site Kayak.com to the sharing widget AddThis—make no such guarantees. Behind the pages you visit, a massive new market for information about what you do online is growing, driven by low-profile but highly profitable personal data companies like BlueKai and Acxiom. Acxiom alone has accumulated an average of 1,500 pieces of data on each person on its database—which includes 96 percent of Americans—along with data about everything from their credit scores to whether they've bought medication for incontinence. And using lightning-fast protocols, any Web site—not just the Googles and Facebooks of the world—can now participate in the fun. In the view of the "behavior market" vendors, every "click signal" you create is a commodity, and every move of your mouse can be auctioned off within microseconds to the highest commercial bidder.

As a business strategy, the Internet giants' formula is simple: The more personally relevant their information offerings are, the more ads they can sell, and the more likely you are to buy the products they're offering. And the formula works. Amazon sells billions of dollars in merchandise by predicting what each customer is interested in and putting it in the front of the virtual store. Up to 60 percent of Netflix's rentals come from the personalized guesses it can make about each customer's movie preferences—and at this point, Netflix can predict how much you'll like a given movie within about half a star. Personalization is a core strategy for the top five sites on the Internet—Yahoo, Google, Facebook, YouTube, and Microsoft Live—as well as countless others.

In the next three to five years, Facebook COO Sheryl Sandberg told one group, the idea of a Web site that isn't customized to a particular user will seem quaint. Yahoo Vice President Tapan Bhat agrees: "The future of the web is about personalization . . . now the web is about 'me.' It's about weaving the web together in a way that is smart and personalized for the user." Google CEO Eric Schmidt enthuses that the "product I've always wanted to build" is Google code that will "guess what I'm trying to type." Google Instant, which guesses what you're searching for as you type and was rolled out in the fall of 2010, is just the start—Schmidt believes that what customers want is for Google to "tell them what they should be doing next."

It would be one thing if all this customization was just about targeted advertising. But personalization isn't just shaping what we buy. For a quickly rising percentage of us, personalized news feeds like Facebook are becoming a primary news source—36 percent of Americans under thirty get their news through social networking sites. And Facebook's popularity is skyrocketing worldwide, with nearly a million more people joining each day. As founder Mark Zuckerberg likes to brag, Facebook may be the biggest source of news in the world (at least for some definitions of "news").

And personalization is shaping how information flows far beyond Facebook, as Web sites from Yahoo News to the *New York Times*-funded startup News.me cater their headlines to our particular interests and desires. It's influencing what videos we watch on YouTube and a dozen smaller competitors, and what blog posts we see. It's affecting whose e-mails we get, which potential mates we run into on OkCupid, and which restaurants are recommended to us on Yelp—which means that personalization could easily have a hand not only in who goes on a date with whom but in where they go and what they talk about. The algorithms that orchestrate our ads are starting to orchestrate our lives.

The basic code at the heart of the new Internet is pretty simple. The new generation of Internet filters looks at the things you seem to like—the actual things you've done, or the things people like you like—and tries to extrapolate. They are prediction engines, constantly creating and refining a theory of who you are and what you'll do and want next. Together, these engines create a unique universe of information for each of us—what I've come to call a filter bubble—which fundamentally alters the way we encounter ideas and information.

Of course, to some extent we've always consumed media that appealed to our interests and avocations and ignored much of the rest. But the filter bubble introduces three dynamics we've never dealt with before.

First, you're alone in it. A cable channel that caters to a narrow interest (say, golf) has other viewers with whom you share a frame of reference. But you're the only person in your bubble. In an age when shared information is the bedrock of shared experience, the filter bubble is a centrifugal force, pulling us apart.

Second, the filter bubble is invisible. Most viewers of conservative or liberal news sources know that they're going to a station curated to serve a particular political viewpoint. But Google's agenda is opaque. Google doesn't tell you who it thinks you are or why it's showing you the results you're seeing. You don't know if its assumptions about you are right or wrong—and you might not even know it's making assumptions about you in the first place. My friend who got more investment-oriented information about BP still has no idea why that was the case— she's not a stockbroker. Because you haven't chosen the criteria by which sites filter information in and out, it's easy to imagine that the information that comes through a filter bubble is unbiased, objective, true. But it's not. In fact, from within the bubble, it's nearly impossible to see how biased it is.

Finally, you don't choose to enter the bubble. When you turn on Fox News or read *The Nation*, you're making a decision about what kind of filter to use to make sense of the world. It's an active process, and like putting on a pair of tinted glasses, you can guess how the editors' leaning shapes your perception. You don't make the same kind of choice with personalized filters. They come to you—and because they drive up profits for the Web sites that use them, they'll become harder and harder to avoid.

OF COURSE, THERE'S a good reason why personalized filters have such a powerful allure. We are overwhelmed by a torrent of information: 900,000 blog posts, 50 million tweets, more than 60 million Facebook status updates, and 210 billion e-mails are sent off into the electronic ether every day. Eric Schmidt likes to point out that if you recorded all human communication from the dawn of time to 2003, it'd take up about 5 billion gigabytes of storage space. Now we're creating that much data every two *days*.

Even the pros are struggling to keep up. The National Security Agency, which copies a lot of the Internet traffic that flows through AT&T's main hub in San Francisco, is building two new stadium-size complexes in the Southwest to process all that data. The biggest problem they face is a lack of power: There literally isn't enough electricity on the grid to support that much computing. The NSA is asking Congress for funds to build new power plants. By 2014, they anticipate dealing with so much data they've invented new units of measurement just to describe it.

Inevitably, this gives rise to what blogger and media analyst Steve Rubel calls the attention crash. As the cost of communicating over large distances and to large groups of people has plummeted, we're increasingly unable to attend to it all. Our focus flickers from text message to Web clip to e-mail. Scanning the ever-widening torrent for the precious bits that are actually important or even just relevant is itself a full-time job.

So when personalized filters offer a hand, we're inclined to take it. In theory, anyway, they can help us find the information we need to know and see and hear, the stuff that really matters among the cat pictures and Viagra ads and treadmill-dancing music videos. Netflix helps you find the right movie to watch in its vast catalog of 140,000 flicks. The Genius function of iTunes calls new hits by your favorite band to your attention when they'd otherwise be lost.

Ultimately, the proponents of personalization offer a vision of a custom-tailored world, every facet of which fits us perfectly. It's a cozy place, populated by our favorite people and things and ideas. If we never want to hear about reality TV (or a more serious issue like gun violence) again, we don't have to—and if we want to hear about every movement of Reese Witherspoon, we can. If we never click on the articles about cooking, or gadgets, or the world outside our country's borders, they simply fade away. We're never bored. We're never annoyed. Our media is a perfect reflection of our interests and desires.

By definition, it's an appealing prospect—a return to a Ptolemaic universe in which the sun and everything else revolves around us. But it comes at a cost: Making everything more personal, we may lose some of the traits that made the Internet so appealing to begin with.

When I began the research that led to the writing of this book, personalization seemed like a subtle, even inconsequential shift. But when I considered what it might mean for a whole society to be adjusted in this way, it started to look more important. Though I follow tech developments pretty closely, I realized there was a lot I didn't know: How did personalization work? What was driving it? Where was it headed? And most important, what will it do to us? How will it change our lives?

In the process of trying to answer these questions, I've talked to sociologists and salespeople, software engineers and law professors. I interviewed one of the founders of OkCupid, an algorithmically driven dating Web site, and one of the chief visionaries of the U.S. information warfare bureau. I learned more than I ever wanted to know about the mechanics of online ad sales and search engines. I argued with cyberskeptics and cybervisionaries (and a few people who were both).

Throughout my investigation, I was struck by the lengths one has to go to in order to fully see what personalization and filter bubbles do. When I interviewed Jonathan McPhie, Google's point man on search personalization, he suggested that it was nearly impossible to guess how the algorithms would shape the experience of any given user. There were simply too many variables and inputs to track. So while Google can look at overall clicks, it's much harder to say how it's working for any one person.

I was also struck by the degree to which personalization is already upon us—not only on Facebook and Google, but on almost every major site on the Web. "I don't think the genie goes back in the bottle," Danny Sullivan told me. Though concerns about personalized media have been raised for a decade— legal scholar Cass Sunstein wrote a smart and provocative book on the topic in 2000—the theory is now rapidly becoming practice: Personalization is already much more a part of our daily experience than many of us realize. We can now begin to see how the filter bubble is actually working, where it's falling short, and what that means for our daily lives and our society.

Every technology has an interface, Stanford law professor Ryan Calo told me, a place where you end and the technology begins. And when the technology's job is to show you the world, it ends up sitting between you and reality, like a camera lens. That's a powerful position, Calo says. "There are lots of ways for it to skew your perception of the world." And that's precisely what the filter bubble does.

THE FILTER BUBBLE'S costs are both personal and cultural. There are direct consequences for those of us who use personalized filters (and soon enough, most of us will, whether we realize it or not). And there are societal consequences, which emerge when masses of people begin to live a filter-bubbled life.

One of the best ways to understand how filters shape our individual experience is to think in terms of our information diet. As sociologist danah boyd said in a speech at the 2009 Web 2.0 Expo:

> Our bodies are programmed to consume fat and sugars because they're rare in nature. . . . In the same way, we're biologically programmed to be attentive to things that stimulate: content that is gross, violent, or sexual and that gossip which is humiliating, embarrassing, or offensive. If we're not careful, we're going to develop the psychological equivalent of obesity. We'll find ourselves consuming content that is least beneficial for ourselves or society as a whole.

Just as the factory farming system that produces and delivers our food shapes what we eat, the dynamics of our media shape what information we consume. Now we're quickly shifting toward a regimen chock-full of personally relevant information. And while that can be helpful, too much of a good thing

can also cause real problems. Left to their own devices, personalization filters serve up a kind of invisible autopropaganda,
indoctrinating us with our own ideas, amplifying our desire for
things that are familiar and leaving us oblivious to the dangers
lurking in the dark territory of the unknown.

In the filter bubble, there's less room for the chance encounters that bring insight and learning. Creativity is often sparked
by the collision of ideas from different disciplines and cultures.
Combine an understanding of cooking and physics and you get
the nonstick pan and the induction stovetop. But if Amazon
thinks I'm interested in cookbooks, it's not very likely to show
me books about metallurgy. It's not just serendipity that's at
risk. By definition, a world constructed from the familiar is a
world in which there's nothing to learn. If personalization is
too acute, it could prevent us from coming into contact with
the mind-blowing, preconception-shattering experiences and
ideas that change how we think about the world and ourselves.

And while the premise of personalization is that it provides
you with a service, you're not the only person with a vested
interest in your data. Researchers at the University of Minne-
sota recently discovered that women who are ovulating respond
better to pitches for clingy clothes and suggested that marketers "strategically time" their online solicitations. With enough
data, guessing this timing may be easier than you think.

At best, if a company knows which articles you read or what
mood you're in, it can serve up ads related to your interests. But
at worst, it can make decisions on that basis that negatively
affect your life. After you visit a page about Third World

backpacking, an insurance company with access to your Web history might decide to increase your premium, law professor Jonathan Zittrain suggests. Parents who purchased EchoMetrix's Sentry software to track their kids online were outraged when they found that the company was then selling their kids' data to third-party marketing firms.

Personalization is based on a bargain. In exchange for the service of filtering, you hand large companies an enormous amount of data about your daily life—much of which you might not trust friends with. These companies are getting better at drawing on this data to make decisions every day. But the trust we place in them to handle it with care is not always warranted, and when decisions are made on the basis of this data that affect you negatively, they're usually not revealed.

Ultimately, the filter bubble can affect your ability to choose how you want to live. To be the author of your life, professor Yochai Benkler argues, you have to be aware of a diverse array of options and lifestyles. When you enter a filter bubble, you're letting the companies that construct it choose which options you're aware of. You may think you're the captain of your own destiny, but personalization can lead you down a road to a kind of informational determinism in which what you've clicked on in the past determines what you see next—a Web history you're doomed to repeat. You can get stuck in a static, ever-narrowing version of yourself—an endless you-loop.

And there are broader consequences. In *Bowling Alone*, his bestselling book on the decline of civic life in America, Robert Putnam looked at the problem of the major decrease in "social capital"—the bonds of trust and allegiance that encourage people to do each other favors, work together to solve common problems, and collaborate. Putnam identified two kinds of social capital: There's the in-group-oriented "bonding" capital created when you attend a meeting of your college alumni, and then there's "bridging" capital, which is created at an event like a town meeting when people from lots of different backgrounds come together to meet each other. Bridging capital is potent: Build more of it, and you're more likely to be able to find that next job or an investor for your small business, because it allows you to tap into lots of different networks for help.

Everybody expected the Internet to be a huge source of bridging capital. Writing at the height of the dot-com bubble, Tom Friedman declared that the Internet would "make us all next door neighbors." In fact, this idea was the core of his thesis in *The Lexus and the Olive Tree*: "The Internet is going to be like a huge vise that takes the globalization system . . . and keeps tightening and tightening that system around everyone, in ways that will only make the world smaller and smaller and faster and faster with each passing day."

Friedman seemed to have in mind a kind of global village in which kids in Africa and executives in New York would build a community together. But that's not what's happening: Our virtual next-door neighbors look more and more like our real-world neighbors, and our real-world neighbors look more and more like us. We're getting a lot of bonding but very little bridging. And this is important because it's bridging that creates our sense of the "public"—the space where we address the problems that transcend our niches and narrow self-interests.

We are predisposed to respond to a pretty narrow set of

stimuli—if a piece of news is about sex, power, gossip, violence, celebrity, or humor, we are likely to read it first. This is the content that most easily makes it into the filter bubble. It's easy to push "Like" and increase the visibility of a friend's post about finishing a marathon or an instructional article about how to make onion soup. It's harder to push the "Like" button on an article titled, "Darfur sees bloodiest month in two years." In a personalized world, important but complex or unpleasant issues— the rising prison population, for example, or homelessness— are less likely to come to our attention at all.

As a consumer, it's hard to argue with blotting out the irrelevant and unlikable. But what is good for consumers is not necessarily good for citizens. What I seem to like may not be what I actually want, let alone what I need to know to be an informed member of my community or country. "It's a civic virtue to be exposed to things that appear to be outside your interest," technology journalist Clive Thompson told me. "In a complex world, almost everything affects you—that closes the loop on pecuniary self-interest." Cultural critic Lee Siegel puts it a different way: "Customers are always right, but people aren't."

THE STRUCTURE OF our media affects the character of our society. The printed word is conducive to democratic argument in a way that laboriously copied scrolls aren't. Television had a profound effect on political life in the twentieth century— from the Kennedy assassination to 9/11—and it's probably not a coincidence that a nation whose denizens spend thirty-six hours a week watching TV has less time for civic life.

The era of personalization is here, and it's upending many of our predictions about what the Internet would do. The creators of the Internet envisioned something bigger and more important than a global system for sharing pictures of pets. The manifesto that helped launch the Electronic Frontier Foundation in the early nineties championed a "civilization of Mind in cyberspace"—a kind of worldwide metabrain. But personalized filters sever the synapses in that brain. Without knowing it, we may be giving ourselves a kind of global lobotomy instead.

From megacities to nanotech, we're creating a global society whose complexity has passed the limits of individual comprehension. The problems we'll face in the next twenty years—energy shortages, terrorism, climate change, and disease—are enormous in scope. They're problems that we can only solve together.

Early Internet enthusiasts like Web creator Tim Berners-Lee hoped it would be a new platform for tackling those problems. I believe it still can be—and as you read on, I'll explain how. But first we need to pull back the curtain—to understand the forces that are taking the Internet in its current, personalized direction. We need to lay bare the bugs in the code—and the coders—that brought personalization to us.

If "code is law," as Larry Lessig famously declared, it's important to understand what the new lawmakers are trying to do. We need to understand what the programmers at Google and Facebook believe in. We need to understand the economic and social forces that are driving personalization, some of which are inevitable and some of which are not. And we need to understand what all this means for our politics, our culture, and our future.

Without sitting down next to a friend, it's hard to tell how the version of Google or Yahoo News that you're seeing differs from anyone else's. But because the filter bubble distorts our perception of what's important, true, and real, it's critically important to render it visible. That is what this book seeks to do.

The Race for Relevance

If you're not paying for something, you're not the customer; you're the product being sold.

—*Andrew Lewis*, under the alias Blue_beetle,
on the Web site MetaFilter

In the spring of 1994, Nicholas Negroponte sat writing and thinking. At the MIT Media Lab, Negroponte's brainchild, young chip designers and virtual-reality artists and robot-wranglers were furiously at work building the toys and tools of the future. But Negroponte was mulling over a simpler problem, one that millions of people pondered every day: what to watch on TV.

By the mid-1990s, there were hundreds of channels streaming out live programming twenty-four hours a day, seven days a week. Most of the programming was horrendous and boring: infomercials for new kitchen gadgets, music videos for the latest one-hit-wonder band, cartoons, and celebrity news. For any given viewer, only a tiny percentage of it was likely to be interesting.

As the number of channels increased, the standard method of surfing through them was getting more and more hopeless.

It's one thing to search through five channels. It's another to search through five hundred. And when the number hits five thousand—well, the method's useless.

But Negroponte wasn't worried. All was not lost: in fact, a solution was just around the corner. "The key to the future of television," he wrote, "is to stop thinking about television as television," and to start thinking about it as a device with embedded intelligence. What consumers needed was a remote control that controls itself, an intelligent automated helper that would learn what each viewer watches and capture the programs relevant to him or her. "Today's TV set lets you control brightness, volume, and channel," Negroponte typed. "Tomorrow's will allow you to vary sex, violence, and political leaning."

And why stop there? Negroponte imagined a future swarming with intelligent agents to help with problems like the TV one. Like a personal butler at a door, the agents would let in only your favorite shows and topics. "Imagine a future," Negroponte wrote, "in which your interface agent can read every newswire and newspaper and catch every TV and radio broadcast on the planet, and then construct a personalized summary. This kind of newspaper is printed in an edition of one. . . . Call it the Daily Me."

The more he thought about it, the more sense it made. The solution to the information overflow of the digital age was smart, personalized, embedded editors. In fact, these agents didn't have to be limited to television; as he suggested to the editor of the new tech magazine *Wired*, "Intelligent agents are the unequivocal future of computing."

In San Francisco, Jaron Lanier responded to this argument with dismay. Lanier was one of the creators of virtual reality; since the eighties, he'd been tinkering with how to bring computers and people together. But the talk of agents struck him as crazy. "What's got into all of you?" he wrote in a missive to the "Wired-style community" on his Web site. "The idea of 'intelligent agents' is both wrong and evil. . . . The agent question looms as a deciding factor in whether [the Net] will be much better than TV, or much worse."

Lanier was convinced that, because they're not actually people, agents would force actual humans to interact with them in awkward and pixelated ways. "An agent's model of what you are interested in will be a cartoon model, and you will see a cartoon version of the world through the agent's eyes," he wrote.

And there was another problem: The perfect agent would presumably screen out most or all advertising. But since online commerce was driven by advertising, it seemed unlikely that these companies would roll out agents who would do such violence to their bottom line. It was more likely, Lanier wrote, that these agents would have double loyalties—bribable agents. "It's not clear who they're working for."

It was a clear and plangent plea. But though it stirred up some chatter in online newsgroups, it didn't persuade the software giants of this early Internet era. They were convinced by Negroponte's logic: The company that figured out how to sift through the digital haystack for the nuggets of gold would win the future. They could see the attention crash coming, as the information options available to each person rose toward infinity.

If you wanted to cash in, you needed to get people to tune in. And in an attention-scarce world, the best way to do that was to provide content that really spoke to each person's idiosyncratic interests, desires, and needs. In the hallways and data centers of Silicon Valley, there was a new watchword: relevance.

Everyone was rushing to roll out an "intelligent" product. In Redmond, Microsoft released Bob—a whole operating system based on the agent concept, anchored by a strange cartoonish avatar with an uncanny resemblance to Bill Gates. In Cupertino, almost exactly a decade before the iPhone, Apple introduced the Newton, a "personal desktop assistant" whose core selling point was the agent lurking dutifully just under its beige surface.

As it turned out, the new intelligent products bombed. In chat groups and on e-mail lists, there was practically an industry of snark about Bob. Users couldn't stand it. *PC World* named it one of the twenty-five worst tech products of all time. And the Apple Newton didn't do much better: Though the company had invested over $100 million in developing the product, it sold poorly in the first six months of its existence. When you interacted with the intelligent agents of the midnineties, the problem quickly became evident: They just weren't that smart.

Now, a decade and change later, intelligent agents are still nowhere to be seen. It looks as though Negroponte's intelligent-agent revolution failed. We don't wake up and brief an e-butler on our plans and desires for the day.

But that doesn't mean they don't exist. They're just hidden. Personal intelligent agents lie under the surface of every Web site we go to. Every day, they're getting smarter and more powerful, accumulating more information about who we are and what we're interested in. As Lanier predicted, the agents don't work only for us: They also work for software giants like Google, dispatching ads as well as content. Though they may lack Bob's cartoon face, they steer an increasing proportion of our online activity.

In 1995 the race to provide personal relevance was just beginning. More than perhaps any other factor, it's this quest that has shaped the Internet we know today.

The John Irving Problem

Jeff Bezos, the CEO of Amazon.com, was one of the first people to realize that you could harness the power of relevance to make a few billion dollars. Starting in 1994, his vision was to transport online bookselling "back to the days of the small bookseller who got to know you very well and would say things like, 'I know you like John Irving, and guess what, here's this new author, I think he's a lot like John Irving,'" he told a biographer. But how to do that on a mass scale? To Bezos, Amazon needed to be "a sort of a small Artificial Intelligence company," powered by algorithms capable of instantly matching customers and books.

In 1994, as a young computer scientist working for Wall Street firms, Bezos had been hired by a venture capitalist to come up with business ideas for the burgeoning Web space. He worked methodically, making a list of twenty products the team could theoretically sell online—music, clothing, electronics—and then digging into the dynamics of each industry. Books started at the bottom of his list, but when he drew up his final results, he was surprised to find them at the top.

Books were ideal for a few reasons. For starters, the book industry was decentralized; the biggest publisher, Random House, controlled only 10 percent of the market. If one publisher wouldn't sell to him, there would be plenty of others who would. And people wouldn't need as much time to get comfortable with buying books online as they might with other products—a majority of book sales already happened outside of traditional bookstores, and unlike clothes, you didn't need to try them on. But the main reason books seemed attractive was simply the fact that there were so many of them—3 million active titles in 1994, versus three hundred thousand active CDs. A physical bookstore would never be able to inventory all those books, but an online bookstore could.

When he reported this finding to his boss, the investor wasn't interested. Books seemed like a kind of backward industry in an information age. But Bezos couldn't get the idea out of his head. Without a physical limit on the number of books he could stock, he could provide hundreds of thousands more titles than industry giants like Borders or Barnes & Noble, and at the same time, he could create a more intimate and personal experience than the big chains.

Amazon's goal, he decided, would be to enhance the process of discovery: a personalized store that would help readers find books and introduce books to readers. But how?

Bezos started thinking about machine learning. It was a

tough problem, but a group of engineers and scientists had been attacking it at research institutions like MIT and the University of California at Berkeley since the 1950s. They called their field "cybernetics"—a word taken from Plato, who coined it to mean a self-regulating system, like a democracy. For the early cyberneticists, there was nothing more thrilling than building systems that tuned themselves, based on feedback. Over the following decades, they laid the mathematical and theoretical foundations that would guide much of Amazon's growth.

In 1990, a team of researchers at the Xerox Palo Alto Research Center (PARC) applied cybernetic thinking to a new problem. PARC was known for coming up with ideas that were broadly adopted and commercialized by others—the graphical user interface and the mouse, to mention two. And like many cutting-edge technologists at the time, the PARC researchers were early power users of e-mail—they sent and received hundreds of them. E-mail was great, but the downside was quickly obvious. When it costs nothing to send a message to as many people as you like, you can quickly get buried in a flood of useless information.

To keep up with the flow, the PARC team started tinkering with a process they called collaborative filtering, which ran in a program called Tapestry. Tapestry tracked how people reacted to the mass e-mails they received—which items they opened, which ones they responded to, and which they deleted—and then used this information to help order the inbox. E-mails that people had engaged with a lot would move to the top of the list; e-mails that were frequently deleted or unopened would go to the bottom. In essence, collaborative filtering was a time-saver: Instead of having to sift through the pile of e-mail yourself, you could rely on others to help presift the items you'd received.

And of course, you didn't have to use it just for e-mail. Tapestry, its creators wrote, "is designed to handle any incoming stream of electronic documents. Electronic mail is only one example of such a stream: others are newswire stories and Net-News articles."

Tapestry had introduced collaborative filtering to the world, but in 1990, the world wasn't very interested. With only a few million users, the Internet was still a small ecosystem, and there just wasn't much information to sort or much bandwidth to download with. So for years collaborative filtering remained the domain of software researchers and bored college students. If you e-mailed ringo@media.mit.edu in 1994 with some albums you liked, the service would send an e-mail back with other music recommendations and the reviews. "Once an hour," according to the Web site, "the server processes all incoming messages and sends replies as necessary." It was an early precursor to Pandora; it was a personalized music service for a pre-broadband era.

But when Amazon launched in 1995, everything changed. From the start, Amazon was a bookstore with personalization built in. By watching which books people bought and using the collaborative filtering methods pioneered at PARC, Amazon could make recommendations on the fly. ("Oh, you're getting *The Complete Dummy's Guide to Fencing*? How about adding a copy of *Waking Up Blind: Lawsuits over Eye Injury*?") And by tracking which users bought what over time, Amazon could start to see which users' preferences were similar. ("Other people who have similar tastes to yours bought this week's new release, *En Garde!*") The more people bought books from Amazon, the better the personalization got.

In 1997, Amazon had sold books to its first million customers. Six months later, it had served 2 million. And in 2001, it reported its first quarterly net profit—one of the first businesses to prove that there was serious money to be made online.

If Amazon wasn't quite able to create the feeling of a local bookstore, its personalization code nonetheless worked quite well. Amazon executives are tight-lipped about just how much revenue it's brought in, but they often point to the personalization engine as a key part of the company's success.

At Amazon, the push for more user data is never-ending: When you read books on your Kindle, the data about which phrases you highlight, which pages you turn, and whether you read straight through or skip around are all fed back into Amazon's servers and can be used to indicate what books you might like next. When you log in after a day reading Kindle e-books at the beach, Amazon is able to subtly customize its site to appeal to what you've read: If you've spent a lot of time with the latest James Patterson, but only glanced at that new diet guide, you might see more commercial thrillers and fewer health books.

Amazon users have gotten so used to personalization that the site now uses a reverse trick to make some additional cash. Publishers pay for placement in physical bookstores, but they can't buy the opinions of the clerks. But as Lanier predicted, buying off algorithms is easy: Pay enough to Amazon, and your book can be promoted as if by an "objective" recommendation by Amazon's software. For most customers, it's impossible to tell which is which.

Amazon proved that relevance could lead to industry dominance. But it would take two Stanford graduate students to apply the principles of machine learning to the whole world of online information.

Click Signals

As Jeff Bezos's new company was getting off the ground, Larry Page and Sergey Brin, the founders of Google, were busy doing their doctoral research at Stanford. They were aware of Amazon's success—in 1997, the dot-com bubble was in full swing, and Amazon, on paper at least, was worth billions. Page and Brin were math whizzes; Page, especially, was obsessed with AI. But they were interested in a different problem. Instead of using algorithms to figure out how to sell products more effectively, what if you could use them to sort through sites on the Web?

Page had come up with a novel approach, and with a geeky predilection for puns, he called it PageRank. Most Web search companies at the time sorted pages using keywords and were very poor at figuring out which page for a given word was the most relevant. In a 1997 paper, Brin and Page dryly pointed out that three of the four major search engines couldn't find themselves. "We want our notion of 'relevant' to only include the very best documents," they wrote, "since there may be tens of thousands of slightly relevant documents."

Page had realized that packed into the linked structure of the Web was a lot more data than most search engines made use of. The fact that a Web page linked to another page could be considered a "vote" for that page. At Stanford, Page had seen professors count how many times their papers had been cited as a rough index of how important they were. Like academic papers, he realized, the pages that a lot of other pages cite—say, the front page of Yahoo—could be assumed to be more "important," and the pages that those pages voted for would matter more. The process, Page argued, "utilized the uniquely democratic structure of the web."

In those early days, Google lived at google.stanford.edu, and Brin and Page were convinced it should be nonprofit and advertising free. "We expect that advertising funded search engines will be inherently biased towards the advertisers and away from the needs of the consumers," they wrote. "The better the search engine is, the fewer advertisements will be needed for the consumer to find what they want. . . . We believe the issue of advertising causes enough mixed incentives that it is crucial to have a competitive search engine that is transparent and in the academic realm."

But when they released the beta site into the wild, the traffic chart went vertical. Google worked—out of the box, it was the best search site on the Internet. Soon, the temptation to spin it off as a business was too great for the twenty-something cofounders to bear.

In the Google mythology, it is PageRank that drove the

company to worldwide dominance. I suspect the company likes it that way—it's a simple, clear story that hangs the search giant's success on a single ingenious breakthrough by one of its founders. But from the beginning, PageRank was just a small part of the Google project. What Brin and Page had really figured out was this: The key to relevance, the solution to sorting through the mass of data on the Web was . . . more data.

It wasn't just which pages linked to which that Brin and Page were interested in. The position of a link on the page, the size of the link, the age of the page—all of these factors mattered. Over the years, Google has come to call these clues embedded in the data *signals*.

From the beginning, Page and Brin realized that some of the most important signals would come from the search engine's users. If someone searches for "Larry Page," say, and clicks on the second link, that's another kind of vote: It suggests that the second link is more relevant to that searcher than the first one. They called this a *click signal*. "Some of the most interesting research," Page and Brin wrote, "will involve leveraging the vast amount of usage data that is available from modern web systems. . . . It is very difficult to get this data, mainly because it is considered commercially valuable." Soon they'd be sitting on one of the world's largest stores of it.

Where data was concerned, Google was voracious. Brin and Page were determined to keep everything: every Web page the search engine had ever landed on, every click every user ever made. Soon its servers contained a nearly real-time copy of most of the Web. By sifting through this data, they were certain they'd find more clues, more signals, that could be used to tweak results. The search-quality division at the company acquired a black-ops kind of feel: few visitors and absolute secrecy were the rule.

"The ultimate search engine," Page was fond of saying, "would understand exactly what you mean and give back exactly what you want." Google didn't want to return thousands of pages of links—it wanted to return one, the one you wanted. But the perfect answer for one person isn't perfect for another. When I search for "panthers," what I probably mean are the large wild cats, whereas a football fan searching for the phrase probably means the Carolina team. To provide perfect relevance, you'd need to know what each of us was interested in. You'd need to know that I'm pretty clueless about football; you'd need to know who I was.

The challenge was getting enough data to figure out what's personally relevant to each user. Understanding what someone means is tricky business—and to do it well, you have to get to know a person's behavior over a sustained period of time.

But how? In 2004, Google came up with an innovative strategy. It started providing other services, services that required users to log in. Gmail, its hugely popular e-mail service, was one of the first to roll out. The press focused on the ads that ran along Gmail's sidebar, but it's unlikely that those ads were the sole motive for launching the service. By getting people to log in, Google got its hands on an enormous pile of data—the hundreds of millions of e-mails Gmail users send and receive each day. And it could cross-reference each user's e-mail and behavior on the site with the links he or she clicked in the Google search engine. Google Apps—a suite of online word-processing and spreadsheet-creation tools—served double duty: It under-cut Microsoft, Google's sworn enemy, and it provided yet another hook for people to stay logged in and continue sending click signals. All this data allowed Google to accelerate the pro-cess of building a theory of identity for each user—what topics each user was interested in, what links each person clicked.

By November 2008, Google had several patents for person-alization algorithms—code that could figure out the groups to which an individual belongs and tailor his or her result to suit that group's preference. The categories Google had in mind were pretty narrow: to illustrate its example in the patent, Google used the example of "all persons interested in collect-ing ancient shark teeth" and "all persons not interested in collecting ancient shark teeth." People in the former category who searched for, say, "Great White incisors" would get differ-ent results from the latter.

Today, Google monitors every signal about us it can get its hands on. The power of this data can't be underestimated: If Google sees that I log on first from New York, then from San Francisco, then from New York again, it knows I'm a bicoastal traveler and can adjust its results accordingly. By looking at what browser I use, it can make some guesses about my age and even perhaps my politics.

How much time you take between the moment you enter your query and the moment you click on a result sheds light on your personality. And of course, the terms you search for reveal a tremendous amount about your interests.

Even if you're not logged in, Google is personalizing your search. The neighborhood—even the block—that you're logging in from is available to Google, and it says a lot about who you are and what you're interested in. A query for "Sox" coming from Wall Street is probably shorthand for the financial legislation "Sarbanes Oxley," while across the Upper Bay in Staten Island it's probably about baseball.

"People always make the assumption that we're done with search," said founder Page in 2009. "That's very far from the case. We're probably only 5 percent of the way there. We want to create the ultimate search engine that can understand anything. . . . Some people could call that artificial intelligence."

In 2006, at an event called Google Press Day, CEO Eric Schmidt laid out Google's five-year plan. One day, he said, Google would be able to answer questions such as "Which college should I go to?" "It will be some years before we can at least partially answer those questions. But the eventual outcome is . . . that Google can answer a more hypothetical question."

Facebook Everywhere

Google's algorithms were unparalleled, but the challenge was to coax users into revealing their tastes and interests. In February 2004, working out of his Harvard dorm room, Mark Zuckerberg came up with an easier approach. Rather than sifting through click signals to figure out what people cared about, the plan behind his creation, Facebook, was to just flat out ask them.

Since he was a college freshman, Zuckerberg had been

interested in what he called the "social graph"—the set of each person's relationships. Feed a computer that data, and it could start to do some pretty interesting and useful things—telling you what your friends were up to, where they were, and what they were interested in. It also had implications for news: In its earliest incarnation as a Harvard-only site, Facebook automatically annotated people's personal pages with links to the *Crimson* articles in which they appeared.

Facebook was hardly the first social network: As Zuckerberg was hacking together his creation in the wee hours of the morning, a hairy, music-driven site named MySpace was soaring; before MySpace, Friendster had for a brief moment captured the attention of the technorati. But the Web site Zuckerberg had in mind was different. It wouldn't be a coy dating site, like Friendster. And unlike MySpace, which encouraged people to connect whether they knew each other or not, Facebook was about taking advantage of existing real-world social connections. Compared to its predecessors, Facebook was stripped down: the emphasis was on information, not flashy graphics or a cultural vibe. "We're a utility," Zuckerberg said later. Facebook was less like a nightclub than a phone company, a neutral platform for communication and collaboration.

Even in its first incarnation, the site grew like wildfire. After Facebook expanded to a few select Ivy League campuses, Zuckerberg's inbox was flooded with requests from students on other campuses, begging him to turn on Facebook for them. By May of 2005, the site was up and running at over eight hundred colleges. But it was the development of the News Feed the following September that pushed Facebook into another league.

On Friendster and MySpace, to find out what your friends were up to, you had to visit their pages. The News Feed algorithm pulled all of these updates out of Facebook's massive database and placed them in one place, up front, right when you logged in. Overnight, Facebook had turned itself from a network of connected Web pages into a personalized newspaper featuring (and created by) your friends. It's hard to imagine a purer source of relevance.

And it was a gusher. In 2006, Facebook users posted literally billions of updates—philosophical quotes, tidbits about who they were dating, what was for breakfast. Zuckerberg and his team egged them on: The more data users handed over to the company, the better their experience could be and the more they'd keep coming back. Early on, they'd added the ability to upload photos, and now Facebook had the largest photo collection in the world. They encouraged users to post links from other Web sites, and millions were submitted. By 2007, Zuckerberg bragged, "We're actually producing more news in a single day for our 19 million users than any other media outlet has in its entire existence."

At first, the News Feed showed nearly everything your friends did on the site. But as the volume of posts and friends increased, the Feed became unreadable and unmanageable. Even if you had only a hundred friends, it was too much to read.

Facebook's solution was EdgeRank, the algorithm that powers the default page on the site, the Top News Feed. EdgeRank ranks every interaction on the site. The math is complicated, but the basic idea is pretty simple, and it rests on three factors.

The first is affinity: The friendlier you are with someone—as determined by the amount of time you spend interacting and checking out his or her profile—the more likely it is that Facebook will show you that person's updates. The second is the relative weight of that type of content: Relationship status updates, for example, are weighted very highly; everybody likes to know who's dating whom. (Many outsiders suspect that the weight, too, is personalized: Different people care about different kinds of content.) The third is time: Recently posted items are weighted over older ones.

EdgeRank demonstrates the paradox at the core of the race for relevancy. To provide relevance, personalization algorithms need data. But the more data there is, the more sophisticated the filters must become to organize it. It's a never-ending cycle.

By 2009, Facebook had hit the 300 million user mark and was growing by 10 million people per month. Zuckerberg, at twenty-five, was a paper billionaire. But the company had bigger ambitions. What the News Feed had done for social information, Zuckerberg wanted to do for all information. Though he never said it, the goal was clear: Leveraging the social graph and the masses of information Facebook's users had provided, Zuckerberg wanted to put Facebook's news algorithm engine at the center of the web.

Even so, it was a surprise when, on April 21, 2010, readers loaded the *Washington Post* homepage and discovered that their friends were on it. In a prominent box in the upper right corner— the place where any editor will tell you the eye lands first—was a feature titled Network News. Each person who visited saw a different set of links in the box—the *Washington Post* links their friends had shared on Facebook. The *Post* was letting Facebook edit its most valuable online asset: its front page. The *New York Times* soon followed suit.

The new feature was one piece of a much bigger rollout, which Facebook called "Facebook Everywhere" and announced at its annual conference, f8 ("fate"). Ever since Steve Jobs sold the Apple by calling it "insanely great," a measure of grandiosity has been part of the Silicon Valley tradition. But when Zuckerberg walked onto the stage on April 21, 2010, his words seemed plausible. "This is the most transformative thing we've ever done for the web," he announced.

The aim of Facebook Everywhere was simple: make the whole Web "social" and bring Facebook-style personalization to millions of sites that currently lack it. Want to know what music your Facebook friends are listening to? Pandora would now tell you. Want to know what restaurants your friends like? Yelp now had the answer. News sites from the *Huffington Post* to the *Washington Post* were now personalized.

Facebook made it possible to press the Like button on any item on the Web. In the first twenty-four hours of the new service, there were 1 billion Likes—and all of that data flowed back into Facebook's servers. Bret Taylor, Facebook's platform lead, announced that users were sharing 25 billion items a month. Google, once the undisputed leader in the push for relevance, seemed worried about the rival a few miles down the road.

The two giants are now in hand-to-hand combat: Facebook poaches key executives from Google; Google's hard at work constructing social software like Facebook. But it's not totally obvious why the two new-media monoliths should be at war. Google, after all, is built around answering questions; Facebook's core mission is to help people connect with their friends.

But both businesses' bottom lines depend on the same thing: targeted, highly relevant advertising. The contextual advertisements Google places next to search results and on Web pages are its only significant source of profits. And while Facebook's finances are private, insiders have made clear that advertising is at the core of the company's revenue model. Google and Facebook have different starting points and different strategies— one starts with the relationships among pieces of information, while the other starts with the relationships among people— but ultimately, they're competing for the same advertising dollars.

From the point of view of the online advertiser, the question is simple. Which company can deliver the most return on a dollar spent? And this is where relevance comes back into the equation. The masses of data Facebook and Google accumulate have two uses. For users, the data provides a key to providing personally relevant news and results. For advertisers, the data is the key to finding likely buyers. The company that has the most data and can put it to the best use gets the advertising dollars.

Which brings us to lock-in. Lock-in is the point at which users are so invested in their technology that even if competitors might offer better services, it's not worth making the switch. If you're a Facebook member, think about what it'd take to get you to switch to another social networking site— even if the site had vastly greater features. It'd probably take a

lot—re-creating your whole profile, uploading all of those pictures, and laboriously entering your friends' names would be
extremely tedious. You're pretty locked in. Likewise, Gmail,
Gchat, Google Voice, Google Docs, and a host of other products are part of an orchestrated campaign for Google lock-in.
The fight between Google and Facebook hinges on which can
achieve lock-in for the most users.

The dynamics of lock-in are described by Metcalfe's law, a
principle coined by Bob Metcalfe, the inventor of the Ethernet
protocol that wires together computers. The law says that the
usefulness of a network increases at an accelerating rate as you
add each new person to it. It's not much use to be the only
person you know with a fax machine, but if everyone you work
with uses one, it's a huge disadvantage not to be in the loop.
Lock-in is the dark side of Metcalfe's law: Facebook is useful in
large part because everyone's on it. It'd take a lot of misman-
agement to overcome that basic fact.

The more locked in users are, the easier it is to convince
them to log in—and when you're constantly logged in, these
companies can keep tracking data on you even when you're not
visiting their Web sites. If you're logged into Gmail and you
visit a Web site that uses Google's Doubleclick ad service, that
fact can be attached to your Google account. And with tracking
cookies these services place on your computer, Facebook or
Google can provide ads based on your personal information on
third-party sites. The whole Web can become a platform for
Google or Facebook.

But Google and Facebook are hardly the only options. The
daily turf warfare between Google and Facebook occupies

scores of business reporters and gigabytes of blog chatter, but there's a stealthy third front opening up in this war. And though most of the companies involved operate under the radar, they may ultimately represent the future of personalization.

The Data Market

The manhunt for accomplices of the 9/11 killers was one of the most extensive in history. In the immediate aftermath of the attacks, the scope of the plot was unclear. Were there more hijackers who hadn't yet been found? How extensive was the network that had pulled off the attacks? For three days, the CIA, FBI, and a host of other acronymed agencies worked around the clock to identify who else was involved. The country's planes were grounded, its airports closed.

When help arrived, it came from an unlikely place. On September 14, the bureau had released the names of the hijackers, and it was now asking—pleading—for anyone with information about the perpetrators to come forward. Later that day, the FBI received a call from Mack McLarty, a former White House official who sat on the board of a little-known but hugely profitable company called Acxiom.

As soon as the hijackers' names had been publicly released, Acxiom had searched its massive data banks, which take up five acres in tiny Conway, Arkansas. And it had found some very interesting data on the perpetrators of the attacks. In fact, it turned out, Acxiom knew more about eleven of the nineteen hijackers than the entire U.S. government did—including their past and current addresses and the names of their housemates.

We may never know what was in the files Acxiom gave the government (though one of the executives told a reporter that Acxiom's information had led to deportations and indictments). But here's what Acxiom knows about 96 percent of American households and half a billion people worldwide: the names of their family members, their current and past addresses, how often they pay their credit card bills whether they own a dog or a cat (and what breed it is), whether they are right-handed or left-handed, what kinds of medication they use (based on pharmacy records) . . . the list of data points is about 1,500 items long.

Acxiom keeps a low profile—it may not be an accident that its name is nearly unpronounceable. But it serves most of the largest companies in America—nine of the ten major credit card companies and consumer brands from Microsoft to Blockbuster. "Think of [Acxiom] as an automated factory," one engineer told a reporter, "where the product we make is data."

To get a sense of Acxiom's vision for the future, consider a travel search site like Travelocity or Kayak. Ever wondered how they make money? Kayak makes money in two ways. One is pretty simple, a holdover from the era of travel agents: When you buy a flight using a link from Kayak, airlines pay the site a small fee for the referral.

The other is much less obvious. When you search for the flight, Kayak places a cookie on your computer—a small file that's basically like putting a sticky note on your forehead saying "Tell me about cheap bicoastal fares." Kayak can then sell that piece of data to a company like Acxiom or its rival Blue-Kai, which auctions it off to the company with the highest bid—in this case, probably a major airline like United. Once it knows what kind of trip you're interested in, United can show you ads for relevant flights—not just on Kayak's site, but on literally almost any Web site you visit across the Internet. This whole process—from the collection of your data to the sale to United—takes under a second.

The champions of this practice call it "behavioral retargeting." Retailers noticed that 98 percent of visitors to online shopping sites leave without buying anything. Retargeting means businesses no longer have to take "no" for an answer.

Say you check out a pair of running sneakers online but leave the site without springing for them. If the shoe site you were looking at uses retargeting, their ads—maybe displaying a picture of the exact sneaker you were just considering—will follow you around the Internet, showing up next to the scores from last night's game or posts on your favorite blog. And if you finally break down and buy the sneakers? Well, the shoe site can sell that piece of information to BlueKai to auction it off to, say, an athletic apparel site. Pretty soon you'll be seeing ads all over the Internet for sweat-wicking socks.

This kind of persistent, personalized advertising isn't just confined to your computer. Sites like Loopt and Foursquare, which broadcast a user's location from her mobile phone, provide advertisers with opportunities to reach consumers with targeted ads even when they're out and about. Loopt is working on an ad system whereby stores can offer special discounts and promotions to repeat customers on their phones—right as they walk through the door. And if you sit down on an air flight, the ads on your seat-back TV screen may be different from your neighbors'. The airline, after all, knows your name and who you are. And by cross-indexing that personal information with a database like Acxiom's, it can know a whole lot more about you. Why not show you your own ads—or, for that matter, a targeted show that makes you more likely to watch them?

TargusInfo, another of the new firms that processes this sort of information, brags that it "delivers more than 62 billion real-time attributes a year." That's 62 billion points of data about who customers are, what they're doing, and what they want. Another ominously named enterprise, the Rubicon Project, claims that its database includes more than half a billion Internet users.

For now, retargeting is being used by advertisers, but there's no reason to expect that publishers and content providers won't get in on it. After all, if the *Los Angeles Times* knows that you're a fan of Perez Hilton, it can front-page its interview with him in your edition, which means you'll be more likely to stay on the site and click around.

What all of this means is that your behavior is now a commodity, a tiny piece of a market that provides a platform for the personalization of the whole Internet. We're used to thinking of the Web as a series of one-to-one relationships: You manage your relationship with Yahoo separately from your relationship with your favorite blog. But behind the scenes, the Web is becoming increasingly integrated. Businesses are realizing that it's profitable to share data. Thanks to Acxiom and the data market, sites can put the most relevant products up front and whisper to each other behind your back.

The push for relevance gave rise to today's Internet giants, and it is motivating businesses to accumulate ever more data about us and to invisibly tailor our online experiences on that basis. It's changing the fabric of the Web. But as we'll see, the consequences of personalization for how we consume news, make political decisions, and even how we think will be even more dramatic.

The User Is the Content

Everything which bars freedom and fullness of communication sets up barriers that divide human beings into sets and cliques, into antagonistic sects and factions, and thereby undermines the democratic way of life.

—*John Dewey*

The technology will be so good, it will be very hard for people to watch or consume something that has not in some sense been tailored for them.

—*Eric Schmidt*, Google CEO

Microsoft Building 1 in Mountain View, California, is a long, low, gunmetal gray hangar, and if it weren't for the cars buzzing by behind it on Highway 101, you'd almost be able to hear the whine of ultrasonic security. On this Saturday in 2010, the vast expanses of parking lot were empty except for a few dozen BMWs and Volvos. A cluster of scrubby pine trees bent in the gusty wind.

Inside, the concrete-floored hallways were crawling with CEOs in jeans and blazers trading business cards over coffee and swapping stories about deals. Most hadn't come far; the startups they represented were based nearby. Hovering over the cheese spread was a group of executives from data firms like Acxiom and Experian who had flown in from Arkansas and New York the night before. With fewer than a hundred people in attendance, the Social Graph Symposium nonetheless included the leaders and luminaries of the targeted-marketing field.

A bell rang, the group filed into breakout rooms, and one of the conversations quickly turned to the battle to "monetize content." The picture, the group agreed, didn't look good for newspapers.

The contours of the situation were clear to anyone paying attention: The Internet had delivered a number of mortal blows to the newspaper business model, any one of which might be fatal. Craigslist had made classified advertisements free, and $18 billion in revenue went *poof.* Nor was online advertising picking up the slack. An advertising pioneer once famously said, "Half the money I spend on advertising is wasted—I just don't know which half." But the Internet turned that logic on its head—with click-through rates and other metrics, businesses suddenly knew exactly which half of their money went to waste. And when ads didn't work as well as the industry had promised, advertising budgets were cut accordingly. Meanwhile, bloggers and freelance journalists started to package and produce news content for free, which pressured the papers to do the same online.

But what most interested the crowd in the room was the fact that the entire premise on which the news business had been built was changing, and the publishers weren't even paying attention.

The *New York Times* had traditionally been able to command high ad rates because advertisers knew it attracted a premium audience—the wealthy opinion-making elite of New York and beyond. In fact, the publisher had a near monopoly on reaching that group—there were only a few other outlets that provided a direct feed into their homes (and out of their pocketbooks).

Now all that was changing. One executive in the marketing session was especially blunt. "The publishers are losing," he said, "and they will lose, because they just don't get it."

Instead of taking out expensive advertisements in the *New York Times*, it was now possible to track that elite cosmopolitan readership using data acquired from Acxiom or BlueKai. This was, to say the least, a game changer in the business of news. Advertisers no longer needed to pay the *New York Times* to reach *Times* readers: they could target them wherever they went online. The era where you had to develop premium content to get premium audiences, in other words, was coming to a close.

The numbers said it all. In 2003, publishers of articles and videos online received most of each dollar advertisers spent on their sites. Now, in 2010, they only received $.20. The difference was moving to the people who had the data—many of whom were in attendance at Mountain View. A PowerPoint presentation circulating in the industry called out the significance of this change succinctly, describing how "premium publishers [were] losing a key advantage" because advertisers can now target premium audiences in "other, cheaper places." The take-home message was clear: Users, not sites, were now the focus.

Unless newspapers could think of themselves as behavioral data companies with a mission of churning out information about their readers' preferences—unless, in other words, they could adapt themselves to the personalized, filter-bubble world—they were sunk.

NEWS SHAPES OUR sense of the world, of what's important, of the scale and color and character of our problems. More important, it provides the foundation of shared experience and shared knowledge on which democracy is built. Unless we understand the big problems our societies face, we can't act together to fix them. Walter Lippmann, the father of modern journalism, put it more eloquently: "All that the sharpest critics of democracy have alleged is true, if there is no steady supply of trustworthy and relevant news. Incompetence and aimlessness, corruption and disloyalty, panic and ultimate disaster must come to any people which is denied an assured access to the facts."

If news matters, newspapers matter, because their journalists write most of it. Although the majority of Americans get their news from local and national TV broadcasts, most of the actual reporting and story generation happens in newspaper newsrooms. They're the core creators of the news economy. Even in 2010, blogs remain incredibly reliant on them: according to Pew Research Center's Project for Excellence in Journalism,

99 percent of the stories linked to in blog posts come from newspapers and broadcast networks, and the *New York Times* and *Washington Post* alone account for nearly 50 percent of all blog links. While rising in importance and influence, net-native media still mostly lack the capacity to shape public life that these papers and a few other outlets like the BBC and CNN have.

But the shift is coming. The forces unleashed by the Internet are driving a radical transformation in who produces news and how they do it. Whereas once you had to buy the whole paper to get the sports section, now you can go to a sports-only Web site with enough new content each day to fill ten papers. Whereas once only those who could buy ink by the barrel could reach an audience of millions, now anyone with a laptop and a fresh idea can.

If we look carefully, we can begin to project the outline of the new constellation that's emerging. This much we know:

- The cost of producing and distributing media of all kinds— words, images, video, and audio streams—will continue to fall closer and closer to zero.
- As a result, we'll be deluged with choices of what to pay attention to—and we'll continue to suffer from "attention crash." This makes curators all the more important. We'll rely ever more heavily on human and software curators to determine what news we should consume.
- Professional human editors are expensive, and code is cheap. Increasingly, we'll rely on a mix of nonprofessional editors (our friends and colleagues) and software code to figure out

what to watch, read, and see. This code will draw heavily on the power of personalization and displace professional human editors.

Many Internet watchers (myself included) cheered the development of "people-powered news"—a more democratic, participatory form of cultural storytelling. But the future may be more machine-powered than people-powered. And many of the breakthrough champions of the people-powered viewpoint tell us more about our current, transitional reality than the news of the future. The story of "Rathergate" is a classic example of the problem.

When CBS News announced nine weeks before the 2004 election that it had papers proving that President Bush had manipulated his military record, the assertion seemed as though it might be the turning point for the Kerry campaign, which had been running behind in the polls. The viewership for *60 Minutes Wednesday* was high. "Tonight, we have new documents and new information on the President's military service and the first-ever interview with the man who says he pulled the strings to get young George W. Bush into the Texas Air National Guard," Dan Rather said somberly as he laid out the facts.

That night, as the *New York Times* was preparing its headline on the story, a lawyer and conservative activist named Harry MacDougald posted to a right-wing forum called Freerepublic .com. After looking closely at the typeface of the documents, MacDougald was convinced that there was something fishy going on. He didn't beat around the bush: "I am saying these documents are forgeries, run through a copier for 15 generations to make them look old," he wrote. "This should be pursued aggressively."

MacDougald's post quickly attracted attention, and the discussion about the forgeries jumped to two other blog communities, *Powerline* and *Little Green Footballs*, where readers quickly discovered other anachronistic quirks. By the next afternoon, the influential *Drudge Report* had the campaign reporters talking about the validity of the documents. And the following day, September 10, the Associated Press, *New York Times, Washington Post*, and other outlets all carried the story: CBS's scoop might not be true. By September 20, the president of CBS News had issued a statement on the documents: "Based on what we now know, CBS News cannot prove that the documents are authentic. . . . We should not have used them." While the full truth of Bush's military record never came to light, Rather, one of the most prominent journalists in the world, retired in disgrace the next year.

Rathergate is now an enduring part of the mythology about the way blogs and the Internet have changed the game of journalism. No matter where you stand on the politics involved, it's an inspiring tale: MacDougald, an activist on a home computer, discovered the truth, took down one of the biggest figures in journalism, and changed the course of an election.

But this version of the story omits a critical point.

In the twelve days between CBS's airing of the story and its public acknowledgment that the documents were probably fakes, the rest of the broadcast news media turned out reams of reportage. The Associated Press and *USA Today* hired professional document reviewers who scrutinized every dot and character. Cable news networks issued breathless updates. A striking 65 percent of Americans—and nearly 100 percent of the political and reportorial classes—were paying attention to the story.

It is only because these news sources reached many of the same people who watch CBS News that CBS could not afford to ignore the story. MacDougald and his allies may have lit the match, but it took print and broadcast media to fan the flames into a career-burning conflagration.

Rathergate, in other words, is a good story about how online and broadcast media can interact. But it tells us little or nothing about how news will move once the broadcast era is fully over—and we're moving toward that moment at a breakneck pace. The question we have to ask is, What does news look like in the postbroadcast world? How does it move? And what impact does it have?

If the power to shape news rests in the hands of bits of code, not professional human editors, is the code up to the task? If the news environment becomes so fragmented that MacDougald's discovery can't reach a broad audience, could Rathergate even happen at all?

Before we can answer that question, it's worth quickly reviewing where our current news system came from.

The Rise and Fall of the General Audience

Lippmann, in 1920, wrote that "the crisis in western democracy is a crisis in journalism." The two are inextricably linked, and to understand the future of this relationship, we have to understand its past.

It's hard to imagine that there was a time when "public opinion" didn't exist. But as late as the mid-1700s, politics was palace politics. Newspapers confined themselves to commercial and foreign news—a report from a frigate in Brussels and a letter from a nobleman in Vienna set in type and sold to the commercial classes of London. Only when the modern, complex, centralized state emerged—with private individuals rich enough to lend money to the king—did forward-looking officials realize that the views of the people outside the walls had begun to matter.

The rise of the public realm—and news as its medium—was partly driven by the emergence of new, complex societal problems, from the transport of water to the challenges of empire, that transcended the narrow bounds of individual experience. But technological changes also made an impact. After all, how news is conveyed profoundly shapes what is conveyed.

While the spoken word is always directed to a specific audience, the written word—and especially the printing press—changed all that. In a real sense, it made the general audience possible. This ability to address a broad, anonymous group fueled the Enlightenment era, and thanks to the printing press, scientists and scholars could spread complex ideas with perfect precision to an audience spread over large distances. And because everyone was literally on the same page, transnational conversations began that would have been impossibly laborious in the earlier scribe-driven epoch.

In the American colonies, the printing industry developed at a fierce clip—at the time of the revolution, there was no other place in the world with such a density and variety of newspapers. And while they catered exclusively to the interests of white male landowners, the newspapers nonetheless provided a common language and common arguments for dissent. Thomas Paine's rallying cry, *Common Sense*, helped give the diverse colonies a sense of mutual interest and solidarity.

Early newspapers existed to provide business owners with information about market prices and conditions, and newspapers depended on subscription and advertising revenues to survive. It wasn't until the 1830s and the rise of the "penny press"—cheap newspapers sold as one-offs on the street—that everyday citizens in the United States became a primary constituency for news. It was at this point that newspapers came to carry what we think of as news today.

The small, aristocratic public was transforming into a general public. The middle class was growing, and because middle-class people had both a day-to-day stake in the life of the nation and the time and money to spend on entertainment, they were hungry for news and spectacle. Circulation skyrocketed. And as education levels went up, more people came to understand the interconnected nature of modern society. If what happened in Russia could affect prices in New York, it was worth following the news from Russia.

But though democracy and the newspaper were becoming ever more intertwined, the relationship wasn't an easy one. After World War I, tensions about what role the newspaper should play boiled over, becoming a matter of great debate among two of the leading intellectual lights of the time, Walter Lippmann and John Dewey.

Lippmann had watched with disgust as newspapers had effectively joined the propaganda effort for World War I. In *Liberty and the News*, a book of essays published in 1921, he angrily assailed the industry. He quoted an editor who had written that in the service of the war, "governments conscripted public opinion. . . . They goose-stepped it. They taught it to stand at attention and salute."

Lippmann wrote that so long as newspapers existed and they determined "by entirely private and unexamined standards, no matter how lofty, what [the average citizen] shall know, and hence what he shall believe, no one will be able to say that the substance of democratic government is secure."

Over the next decade, Lippmann advanced his line of thought. Public opinion, Lippmann concluded, was too malleable—people were easily manipulated and led by false information. In 1925, he wrote *The Phantom Public*, an attempt to dismantle the illusion of a rational, informed populace once and for all. Lippmann argued against the prevailing democratic mythology, in which informed citizens capably made decisions about the major issues of the day. The "omnicompetent citizens" that such a system required were nowhere to be found. At best, ordinary citizens could be trusted to vote out the party that was in power if it was doing too poorly; the real work of governance, Lippmann argued, should be entrusted to insider experts who had education and expertise to see what was really going on.

John Dewey, one of the great philosophers of democracy, couldn't pass up the opportunity to engage. In The Public and Its Problems, a series of lectures Dewey gave in response to Lippmann's book, he admitted that many of Lippmann's

critiques were not wrong. The media were able to easily manipulate what people thought. Citizens were hardly informed
enough to properly govern.

However, Dewey argued, to accept Lippmann's proposal
was to give up on the promise of democracy—an ideal that had
not yet fully been realized but might still be. "To learn to be
human," Dewey argued, "is to develop through the give and
take of communication an effective sense of being an individu-
ally distinctive member of a community." The institutions of
the 1920s, Dewey said, were closed off—they didn't invite
democratic participation. But journalists and newspapers could
play a critical role in this process by calling out the citizen
in people—reminding them of their stake in the nation's
business.

While they disagreed on the contours of the solution, Dewey
and Lippmann did fundamentally agree that news making was
a fundamentally political and ethical enterprise—and that publishers had to handle their immense responsibility with great
care. And because the newspapers of the time were making
money hand over fist, they could afford to listen. At Lippmann's
urging, the more credible papers built a wall between the
business portion of their papers and the reporting side. They
began to champion objectivity and decry tilted reporting. It's
this ethical model—one in which newspapers have a responsibility to both neutrally inform and convene the public—which
guided the aspirations of journalistic endeavors for the last half
century.

Of course, news agencies have frequently fallen short of
these lofty goals—and it's not always clear how hard they even

try. Spectacle and profit seeking frequently win out over good journalistic practice; media empires make reporting decisions to placate advertisers; and not every outlet that proclaims itself "fair and balanced" actually is.

Thanks to critics like Lippmann, the present system has a sense of ethics and public responsibility baked in, however imperfectly. But though it's playing some of the same roles, the filter bubble does not.

A New Middleman

New York Times critic Jon Pareles calls the 2000s the disintermediation decade. *Disintermediation*—the elimination of middlemen—is "the thing that the Internet does to every business, art, and profession that aggregates and repackages," wrote protoblogger Dave Winer in 2005. "The great virtue of the Internet is that it erodes power," says the Internet pioneer Esther Dyson. "It sucks power out of the center, and takes it to the periphery, it erodes the power of institutions over people while giving to individuals the power to run their own lives."

The disintermediation story was repeated hundreds of times, on blogs, in academic papers, and on talk shows. In one familiar version, it goes like this: Once upon a time, newspaper editors woke up in the morning, went to work, and decided what we should think. They could do this because printing presses were expensive, but it became their explicit ethos: As newspapermen, it was their paternalistic duty to feed the citizenry a healthy diet of coverage.

Many of them meant well. But living in New York and Washington, D.C., they were enthralled by the trappings of power. They counted success by the number of insider cocktail parties they were invited to, and the coverage followed suit. The editors and journalists became embedded in the culture they were supposed to cover. And as a result, powerful people got off the hook, and the interests of the media tilted against the interests of everyday folk, who were at their mercy.

Then the Internet came along and disintermediated the news. All of a sudden, you didn't have to rely on the *Washington Post*'s interpretation of the White House press briefing—you could look up the transcript yourself. The middleman dropped out—not just in news, but in music (no more need for *Rolling Stone*—you could now hear directly from your favorite band) and commerce (you could follow the Twitter feed of the shop down the street) and nearly everything else. The future, the story says, is one in which we go direct.

It's a story about efficiency and democracy. Eliminating the evil middleman sitting between us and what we want sounds good. In a way, disintermediation is taking on the idea of media itself. The word *media*, after all, comes from the Latin for "middle layer." It sits between us and the world; the core bargain is that it will connect us to what's happening but at the price of direct experience. Disintermediation suggests we can have both.

There's some truth to the description, of course. But while enthrallment to the gatekeepers is a real problem, disintermediation is as much mythology as fact. Its effect is to make the new mediators—the new gatekeepers—invisible. "It's about the many wresting power from the few," *Time* magazine announced when it made "you" the person of the year. But as law professor and *Master Switch* author Tim Wu says, "The rise of networking did not eliminate intermediaries, but rather changed who they are." And while power moved toward consumers, in the sense that we have exponentially more choice about what media we consume, the power still isn't held by consumers.

Most people who are renting and leasing apartments don't "go direct"—they use the intermediary of craigslist. Readers use Amazon.com. Searchers use Google. Friends use Facebook. And these platforms hold an immense amount of power—as much, in many ways, as the newspaper editors and record labels and other intermediaries that preceded them. But while we've raked the editors of the *New York Times* and the producers of CNN over the coals for the stories they've missed and the interests they've served, we've given very little scrutiny to the interests behind the new curators.

In July 2010, Google News rolled out a personalized version of its popular service. Sensitive to concerns about shared experience, Google made sure to highlight the "top stories" that are of broad, general interest. But look below that top band, and you will see only stories that are locally and personally relevant to you, based on the interests that you've demonstrated through Google and what articles you've clicked on in the past. Google's CEO doesn't beat around the bush when he describes where this is all headed: "Most people will have personalized news-reading experiences on mobile-type devices that will largely replace their traditional reading of newspapers," he tells an interviewer. "And that that kind of news consumption will be very personal, very targeted. It will remember what you know. It will suggest things that you might want to know. It will have advertising. Right? And it will be as convenient and fun as reading a traditional newspaper or magazine."

Since Krishna Bharat created the first prototype of Google News to monitor worldwide coverage after 9/11, Google News has become one of the top global portals for news. Tens of millions of visitors pull up the site each month—more than visit the BBC. Speaking at the IJ-7 Innovation Journalism conference at Stanford—to a room full of fairly anxious newspaper professionals—Bharat laid out his vision: "Journalists," Bharat explained, "should worry about creating the content and other people in technology should worry about bringing the content to the right group—given the article, what's the best set of eyeballs for it, and that can be solved by personalization."

In many ways, Google News is still a hybrid model, driven in part by the judgment of a professional editorial class. When a Finnish editor asked Bharat what determines the priority of stories, he emphasized that newspaper editors themselves still have disproportionate control: "We pay attention," he said, "to the editorial decisions that different editors have made: what your paper chose to cover, when you published it, and where you placed it on your front page." *New York Times* editor Bill Keller, in other words, still has a disproportionate ability to affect a story's prominence on Google News.

It's a tricky balance: On the one hand, Bharat tells an interviewer, Google should promote what the reader enjoys reading. But at the same time, overpersonalization that, for example, excludes important news from the picture would be a disaster.

Bharat doesn't seem to have fully resolved the dilemma, even for himself. "I think people care about what other people care about, what other people are interested in—most important, their social circle," he says.

Bharat's vision is to move Google News off Google's site and onto the sites of other content producers. "Once we get personalization working for news," Bharat tells the conference, "we can take that technology and make it available to publishers, so they can [transform] their website appropriately" to suit the interests of each visitor.

Krishna Bharat is in the hot seat for a good reason. While he's respectful to the front page editors who pepper him with questions, and his algorithm depends on their expertise, Google News, if it's successful, may ultimately put a lot of front-page editors out of work. Why visit your local paper's Web site, after all, if Google's personalized site has already pulled the best pieces?

The Internet's impact on news was explosive in more ways than one. It expanded the news space by force, sweeping older enterprises out of its path. It dismantled the trust that news organizations had built. In its wake lies a more fragmented and shattered public space than the one that came before.

It's no secret that trust in journalists and news providers has plummeted in recent years. But the shape of the curve is mysterious: According to a Pew poll, Americans lost more faith in news agencies between 2007 and 2010 than they did in the prior twelve years. Even the debacle over Iraq's WMDs didn't make much of a dent in the numbers—but whatever happened in 2007 did.

While we still don't have conclusive proof, it appears that this, too, is an effect of the Internet. When you're getting news from one source, the source doesn't draw your attention much to its own errors and omissions. Corrections, after all, are buried in tiny type on an inside page. But as masses of news readers went online and began to hear from multiple sources, the differences in coverage were drawn out and amplified. You don't hear about the *New York Times*'s problems much from the *New York Times*—but you do hear about them from political blogs, like the *Daily Kos* or *Little Green Footballs*, and from groups on both sides of the spectrum, like MoveOn or RightMarch. More voices, in other words, means less trust in any given voice.

As Internet thinker Clay Shirky has pointed out, the new, low trust levels may not be inappropriate. It may be that the broadcast era kept trust artificially high. But as a consequence, for most of us now, the difference in authority between a blog post and an article in the *New Yorker* is much smaller than one would think.

Editors at Yahoo News, the biggest news site on the Internet, can see this trend in action. With over 85 million daily visitors, when Yahoo links to articles on other servers—even those of nationally known papers—it has to give technicians advance warning so that they can handle the load. A single link can generate up to 12 million views. But according to an executive in the news department, it doesn't matter much to Yahoo's users where the news is coming from. A spicy headline will win over a more trusted news source any day. "People don't make much of a distinction between the *New York Times* and some random blogger," the executive told me.

This is Internet news: Each article ascends the most-forwarded lists or dies an ignominious death on its own. In the old days, *Rolling Stone* readers would get the magazine in the mail and leaf through it; now, the popular stories circulate online independent of the magazine. I read the exposé on General Stanley McChrystal but had no idea that the cover story was about Lady Gaga. The attention economy is ripping the binding, and the pages that get read are the pages that are frequently the most topical, scandalous, and viral.

Nor is debundling just about print media. While the journalistic hand-wringing has focused mostly on the fate of the newspaper, TV channels face the same dilemma. From Google to Microsoft to Comcast, executives are quite clear that what they call convergence is coming soon. Close to a million Americans are unplugging from cable TV offerings and getting their video online every year—and those numbers will accelerate as more services like Netflix's movie-on-demand and Hulu go online. When TV goes fully digital, channels become little more than brands—and the order of programs, like the order of articles, is determined by the user's interest and attention, not the station manager.

And of course, that opens the door for personalization. "Internet connected TV is going to be a reality. It will dramatically change the ad industry forever. Ads will become interactive and delivered to individual TV sets according to the user," Google VP for global media Henrique de Castro has said. We may say good-bye, in other words, to the yearly ritual of the Super Bowl commercial, which won't create the same buzz when everyone is watching different ads.

If trust in news agencies is falling, it is rising in the new realm of amateur and algorithmic curation. If the newspaper and magazine are being torn apart on one end, the pages are being recompiled on the other—a different way every time. Facebook is an increasingly vital source of news for this reason: Our friends and family are more likely to know what's important and relevant to us than some newspaper editor in Manhattan.

Personalization proponents often point to social media like Facebook to dispute the notion that we'll end up in a narrow, overfiltered world. Friend your softball buddy on Facebook, the argument goes, and you'll have to listen to his political rants even if you disagree.

Since they have trust, it's true that the people we know can bring some focus to topics outside our immediate purview. But there are two problems with relying on a network of amateur curators. First, by definition, the average person's Facebook friends will be much more like that person than a general-interest news source. This is especially true because our physical communities are becoming more homogeneous as well—and we generally know people who live near us. Because your softball buddy lives near you, he's likely to share many of your views. It's ever less likely that we'll come to be close with people very different from us, online or off—and thus it's less likely we'll come into contact with different points of view.

Second, personalization filters will get better and better at overlaying themselves on individuals' recommendations. Like your friend Sam's posts on football but not his erratic musings on *CSI*? A filter watching and learning which pieces of content you interact with can start to sift one from another—and undermine even the limited leadership that a group of friends and pundits can offer. Google Reader, another product from Google that helps people manage streams of posts from blogs, now has a feature called Sort by Magic, which does precisely this.

This leads to the final way in which the future of media is likely to be different than we expected. Since its early days, Internet evangelists have argued that it was an inherently active medium. "We think basically you watch television to turn your brain off, and you work on your computer when you want to turn your brain on," Apple founder Steve Jobs told *Macworld* in 2004.

Among techies, these two paradigms came to be known as push technology and pull technology. A Web browser is an example of pull technology: You put in an address, and your computer pulls information from that server. Television and the mail, on the other hand, are push technologies: The information shows up on the tube or at your doorstop without any action on your end. Internet enthusiasts were excited about the shift from push to pull for reasons that are now pretty obvious: Rather than wash the masses in waves of watered-down, lowest-common-denominator content, pull media put users in control.

The problem is that pull is actually a lot of work. It requires you to be constantly on your feet, curating your own media experience. That's way more energy than TV requires during the whopping thirty-six hours a week that Americans watch today.

In TV network circles, there's a name for the passive way with which Americans make most of those viewing decisions: the theory of least objectionable programming. Researching TV viewers' behavior in the 1970s, pay-per-view innovator Paul Klein noticed that people quit channel surfing far more quickly than one might suspect. During most of those thirty-six hours a week, the theory suggests, we're not looking for a program in particular. We're just looking to be unobjectionably entertained.

This is part of the reason TV advertising has been such a bonanza for the channel's owners. Because people watch TV passively, they're more likely to keep watching when ads come on. When it comes to persuasion, passive is powerful.

While the broadcast TV era may be coming to a close, the era of least objectionable programming probably isn't—and personalization stands to make the experience even more, well, unobjectionable. One of YouTube's top corporate priorities is the development of a product called LeanBack, which strings together videos in a row to provide the benefits of push and pull. It's less like surfing the Web and more like watching TV—a personalized experience that lets the user do less and less. Like the music service Pandora, LeanBack viewers can easily skip videos and give the viewer feedback for picking the next videos—thumbs up for this one, thumbs down for these three. LeanBack would learn. Over time, the vision is for Lean-Back to be like your own personal TV channel, stringing together content you're interested in while requiring less and less engagement from you.

Steve Jobs's proclamation that computers are for turning

your brain on may have been a bit too optimistic. In reality, as personalized filtering gets better and better, the amount of energy we'll have to devote to choosing what we'd like to see will continue to decrease.

And while personalization is changing our experience of news, it's also changing the economics that determine what stories get produced.

The Big Board

The offices of Gawker Media, the ascendant blog empire based in SoHo, look little like the newsroom of the *New York Times* a few miles to the north. But the driving difference between the two is the flat-screen TV that hovers over the room.

This is the Big Board, and on it are a list of articles and numbers. The numbers represent the number of times each article has been read, and they're big: Gawker's Web sites routinely see hundreds of millions of page views a month. The Big Board captures the top posts across the company's Web sites, which focus on everything from media (Gawker) to gadgets (Gizmodo) to porn (Fleshbot). Write an article that makes it onto the Big Board, and you're liable to get a raise. Stay off it for too long, and you may need to find a different job.

At the *New York Times*, reporters and bloggers aren't allowed to see how many people click on their stories. This isn't just a rule, it's a philosophy that the *Times* lives by: The point of being the newspaper of record is to provide readers with the benefit of excellent, considered editorial judgment. "We don't let metrics dictate our assignments and play," *New York Times* editor Bill Keller said, "because we believe readers come to us for our judgment, not the judgment of the crowd. We're not 'American Idol.'" Readers can vote with their feet by subscribing to another paper if they like, but the *Times* doesn't pander. Younger *Times* writers who are concerned about such things have to essentially bribe the paper's system administrators to give them a peek at their stats. (The paper does use aggregate statistics to determine which online features to expand or cut.)

If the Internet's current structures mostly tend toward fragmentation and local homogeneity, there is one exception: The only thing that's better than providing articles that are relevant to you is providing articles that are relevant to everyone. Traffic watching is a new addiction for bloggers and managers—and as more sites publish their most-popular lists, readers can join in the fun too.

Of course, journalistic traffic chasing isn't exactly a new phenomenon: Since the 1800s, papers have boosted their circulations with sensational reports. Joseph Pulitzer, in honor of whom the eponymous prizes are awarded each year, was a pioneer of using scandal, sex, fearmongering, and innuendo to drive sales.

But the Internet adds a new level of sophistication and granularity to the pursuit. Now the *Huffington Post* can put an article on its front page and know within minutes whether it's trending viral; if it is, the editors can kick it by promoting it more heavily. The dashboard that allows editors to watch how stories are doing is considered the crown jewel of the enterprise. Associated Content pays an army of online contributors small amounts to troll search queries and write pages that answer the most common questions; those whose pages see a lot of traffic share in the advertising revenue. Sites like Digg and Reddit attempt to turn the whole Internet into a most-popular list with increasing sophistication, by allowing users to vote submitted articles from throughout the Web onto the site's front page. Reddit's algorithm even has a kind of physics built into it so that articles that don't receive a constant amount of approval will begin to fade, and its front page mixes the articles the group thinks are most important with your personal preferences and behavior—a marriage of the filter bubble and the most-popular list.

Las Últimas Noticias, a major paper in Chile, began basing its content entirely on what readers clicked on in 2004: Stories with lots of clicks got follow-ups, and stories with no clicks got killed. The reporters don't have beats anymore—they just try to gin up stories that will get clicks.

At Yahoo's popular *Upshot* news blog, a team of editors mine the data produced by streams of search queries to see what terms people are interested in, in real time. Then they produce articles responsive to those queries: When a lot of people search for "Obama's birthday," *Upshot* produces an article in response, and soon the searchers are landing on a Yahoo page and seeing Yahoo advertising. "We feel like the differentiator here, what separates us from a lot of our competitors is our ability to aggregate all this data," the vice president of Yahoo Media told the *New York Times*. "This idea of creating content in response to audience insight and audience needs is one component of the strategy, but it's a big component."

And what tops the traffic charts? "If it bleeds, it leads" is one of the few news maxims that has continued into the new era. Obviously, what's popular differs among audiences: A study of the *Times*'s most-popular list found that articles that touched on Judaism were often forwarded, presumably due to the *Times*'s readership. In addition, the study concluded, "more practically useful, surprising, affect-laden, and positively valenced articles are more likely to be among the newspaper's most e-mailed stories on a given day, as are articles that evoke more awe, anger, and anxiety, and less sadness."

Elsewhere, the items that top most-popular lists get a bit more crass. The site Buzzfeed recently linked to the "headline that has everything" from Britain's *Evening Herald:* "Woman in Sumo Wrestler Suit Assaulted Her Ex-girlfriend in Gay Pub After She Waved at a Man Dressed as a Snickers Bar." The top story in 2005 for the *Seattle Times* stayed on the most-read list for weeks; it concerned a man who died after having sex with a horse. The *Los Angeles Times*'s top story in 2007 was an article about the world's ugliest dog.

Responsiveness to the audience sounds like a good thing— and in a lot of cases, it is. "If we view the role of cultural products as giving us something to talk about," writes a *Wall Street Journal* reporter who looked into the most-popular phenomenon, "then the most important thing might be that everyone sees the same thing and not what the thing is." Traffic chasing takes media making off its Olympian heights, placing journalists and editors on the same plane with everyone else. The *Washington Post* ombudsman described journalists' often paternalistic approach to readers: "In a past era, there was little need to share marketing information with the *Post*'s news-room. Profits were high. Circulation was robust. Editors decided what they thought readers needed, not necessarily what they wanted."

The Gawker model is almost the precise opposite. If the *Washington Post* emulates Dad, these new enterprises are more like fussy, anxious children squalling to be played with and picked up.

When I asked him about the prospects for important but unpopular news, the Media Lab's Nicholas Negroponte smiled. On one end of the spectrum, he said, is sycophantic personalization—"You're so great and wonderful, and I'm going to tell you exactly what you want to hear." On the other end is the parental approach: "I'm going to tell you this whether you want to hear this or not, because you need to know." Currently, we're headed in the sycophantic direction. "There will be a long period of adjustment," says Professor Michael Schudson, "as the separation of church and state is breaking down, so to speak. In moderation, that seems okay, but Gawker's Big Board is a scary extreme, it's surrender."

Of Apple and Afghanistan

Google News pays more attention to political news than many of the creators of the filter bubble. After all, it draws in large part on the decisions of professional editors. But even in Google News, stories about Apple trump stories about the war in Afghanistan.

I enjoy my iPhone and iPad, but it's hard to argue that these things are of similar importance to developments in Afghanistan. But this Apple-centric ranking is indicative of what the combination of popular lists and the filter bubble will leave out: Things that are important but complicated. "If traffic ends up guiding coverage," the *Washington Post*'s ombudsman writes, "will *The Post* choose not to pursue some important stories because they're 'dull'?"

Will an article about, say, child poverty ever seem hugely personally relevant to many of us, beyond the academics studying the field and the people directly affected? Probably not, but it's still important to know about.

Critics on the left frequently argue that the nation's top media underreport the war. But for many of us, myself included, reading about Afghanistan is a chore. The story is convoluted, confusing, complex, and depressing.

In the editorial judgment of the *Times*, however, I need to know about it, and because they persist in putting it on the front page despite what must be abominably low traffic rates, I continue to read about it. (This doesn't mean the *Times* is overruling my own inclinations. It's just supporting one of my inclinations—to be informed about the world—over the more immediate inclination to click on whatever tickles my fancy.) There are places where media that prioritize importance over popularity or personal relevance are useful—even necessary.

Clay Shirky points out that newspaper readers always mostly skipped over the political stuff. But to do so, they had to at least glance at the front page—and so, if there was a huge political scandal, enough people would know about it to make an impact at the polls. "The question," Shirky says, "is how can the average citizen ignore news of the day to the ninety-ninth percentile and periodically be alarmed when there is a crisis? How do you threaten business and civic leaders with the possibility that if things get too corrupt, the alarm can be sounded?" The front page played that role—but now it's possible to skip it entirely.

Which brings us back to John Dewey. In Dewey's vision, it is these issues—"indirect, extensive, enduring and serious consequences of conjoint and interacting behavior"—that call the public into existence. The important matters that indirectly touch all of our lives but exist out of the sphere of our immediate self-interest are the bedrock and the raison d'être of democracy. *American Idol* may unite a lot of us around the same fireplace, but it doesn't call out the citizen in us. For better or worse—I'd argue for better—the editors of the old media did.

There's no going back, of course. Nor should there be: the Internet still has the potential to be a better medium for democracy than broadcast, with its one-direction-only information flows, ever could be. As journalist A. J. Liebling pointed out, freedom of the press was for those who owned one. Now we all do.

But at the moment, we're trading a system with a defined and well-debated sense of its civic responsibilities and roles for one with no sense of ethics. The Big Board is tearing down the wall between editorial decision-making and the business side of the operation. While Google and others are beginning to grapple with the consequences, most personalized filters have no way of prioritizing what really matters but gets fewer clicks. And in the end, "Give the people what they want" is a brittle and shallow civic philosophy.

But the rise of the filter bubble doesn't just affect how we process news. It can also affect how we think.

The Adderall Society

It is hardly possible to overrate the value . . . of placing human beings in contact with persons dissimilar to themselves, and with modes of thought and action unlike those with which they are familiar. . . . Such communication has always been, and is peculiarly in the present age, one of the primary sources of progress.

—*John Stuart Mill*

The manner in which some of the most important individual discoveries were arrived at reminds one more of a sleepwalker's performance than an electronic brain's.

—*Arthur Koestler, The Sleepwalkers*

In the spring of 1963, Geneva was swarming with diplomats. Delegations from eighteen countries had arrived for negotiations on the Nuclear Test Ban treaty, and meetings were under way in scores of locations throughout the Swiss city. After one afternoon of discussions between the American and Russian delegations, a young KGB officer approached a forty-year-old American diplomat named David Mark. "I'm new on the Soviet delegation, and I'd like to talk to you," he whispered to Mark in Russian, "but I don't want to talk here. I want to have lunch with you." After reporting the contact to colleagues at the CIA, Mark agreed, and the two men planned a meeting at a local restaurant the following day.

At the restaurant, the officer, whose name was Yuri Nosenko, explained that he'd gotten into a bit of a scrape. On his first night in Geneva, Nosenko had drunk too much and brought a prostitute back to his hotel room. When he awoke, to his horror, he found that his emergency stash of $900 in Swiss francs was missing—no small sum in 1963. "I've got to make it up," Nosenko told him. "I can give you some information that will be very interesting to the CIA, and all I want is my money." They set up a second meeting, to which Nosenko arrived in an obviously inebriated state. "I was snookered," Nosenko admitted later—"very drunk."

In exchange for the money, Nosenko promised to spy for the CIA in Moscow, and in January 1964 he met directly with CIA handlers to discuss his findings. This time, Nosenko had big news: He claimed to have handled the KGB file of Lee Harvey Oswald and said it contained nothing suggesting the Soviet Union had foreknowledge of Kennedy's assassination, potentially ruling out Soviet involvement in the event. He was willing to share more of the file's details with the CIA if he would be allowed to defect and resettle in the United States.

Nosenko's offer was quickly transmitted to CIA headquarters in Langley, Virginia. It seemed like a potentially enormous break: Only months after Kennedy had been shot, determining who was behind his assassination was one of the agency's top priorities. But how could they know if Nosenko was telling the truth? James Jesus Angleton, one of the lead agents on Nosenko's case, was skeptical. Nosenko could be a trap—even part of a "master plot" to draw the CIA off the trail. After much discussion, the agents agreed to let Nosenko defect: If he was lying, it would indicate that the Soviet Union *did* know something about Oswald, and if he was telling the truth, he would be useful for counterintelligence.

As it turned out, they were wrong about both. Nosenko traveled to the United States in 1964, and the CIA collected a massive, detailed dossier on their latest catch. But almost as soon as he started the debriefing process, inconsistencies began to emerge. Nosenko claimed he'd graduated from his officer training program in 1949, but the CIA's documents indicated otherwise. He claimed to have no access to documents that KGB officers of his station ought to have had. And why was this man with a wife and child at home in Russia defecting without them?

Angleton became more and more suspicious, especially after his drinking buddy Kim Philby was revealed to be a Soviet spy. Clearly, Nosenko was a decoy sent to dispute and undermine the intelligence the agency was getting from another Soviet defector. The debriefings became more intense. In 1964, Nosenko was thrown into solitary confinement, where he was subjected for several years to harsh interrogation intended to break him and force him to confess. In one week, he was subjected to polygraph tests for twenty-eight and a half hours. Still, no break was forthcoming.

Not everyone at the CIA thought Nosenko was a plant. And as more details from his biography became clear, it came to seem more and more likely that the man they had imprisoned was no spymaster. Nosenko's father was the minister of shipbuilding and a member of the Communist Party Central Committee who had buildings named after him. When young Yuri had been caught stealing at the Naval Preparatory School and was beaten up by his classmates, his mother had complained directly to Stalin; some of his classmates were sent to the Russian front as punishment. It was looking more and more as though Yuri was just "the spoiled-brat son of a top leader" and a bit of a mess. The reason for the discrepancy in graduation dates became clear: Nosenko had been held back a year in school for flunking his exam in Marxism-Leninism, and he was ashamed of it.

By 1968, the balance of senior CIA agents came to believe that the agency was torturing an innocent man. They gave him $80,000, and set him up in a new identity somewhere in the American South. But the emotional debate over his veracity continued to haunt the CIA for decades, with "master plan" theorists sparring with those who believed he was telling the truth. In the end, six separate investigations were made into Nosenko's case. When he passed away in 2008, the news of his death was relayed to the *New York Times* by a "senior intelligence official" who refused to be identified.

One of the officials most affected by the internal debate was an intelligence analyst by the name of Richards Heuer. Heuer had been recruited to the CIA during the Korean War, but he had always been interested in philosophy, and especially the branch known as epistemology—the study of knowledge. Although Heuer wasn't directly involved in the Nosenko case, he was required to be briefed on it for other work he was doing, and he'd initially fallen for the "master plot" hypothesis. Years later, Heuer set out to analyze the analysts—to figure out where the flaws were in the logic that had led to Nosenko's lost years in a CIA prison. The result is a slim volume called *The Psychology of Intelligence Analysis*, whose preface is full of laudatory comments by Heuer's colleagues and bosses. The book is a kind of Psychology and Epistemology 101 for would-be spooks.

For Heuer, the core lesson of the Nosenko debacle was clear: "Intelligence analysts should be self-conscious about their reasoning processes. They should think about how they make judgments and reach conclusions, not just about the judgments and conclusions themselves."

Despite evidence to the contrary, Heuer wrote, we have a tendency to believe that the world is as it appears to be. Children eventually learn that a snack removed from view doesn't disappear from the universe, but even as we mature we still tend to conflate seeing with believing. Philosophers call this view naïve realism, and it is as seductive as it is dangerous. We tend to believe we have full command of the facts and that the patterns we see in them are facts as well. (Angleton, the "master theory" proponent, was sure that Nosenko's pattern of factual errors indicated that he was hiding something and was breaking under pressure.)

So what's an intelligence analyst—or anyone who wants to get a good picture of the world, for that matter—to do? First, Heuer suggests, we have to realize that our idea of what's real often comes to us secondhand and in a distorted form—edited, manipulated, and filtered through media, other human beings, and the many distorting elements of the human mind.

Nosenko's case was riddled with these distorting factors, and the unreliability of the primary source was only the most obvious one. As voluminous as the set of data that the CIA had compiled on Nosenko was, it was incomplete in certain important ways: The agency knew a lot about his rank and status but had learned very little about his personal background and internal life. This led to a basic unquestioned assumption: "The KGB would never let a screw-up serve at this high level; therefore, he must be deceiving us."

"To achieve the clearest possible image" of the world, Heuer writes, "analysts need more than information. . . . They also need to understand the lenses through which this information passes." Some of these distorting lenses are outside of our heads. Like a biased sample in an experiment, a lopsided selection of data can create the wrong impression: For a number of structural and historical reasons, the CIA record on Nosenko was woefully inadequate when it came to the man's personal history. And some of them are cognitive processes: We tend to convert "lots of pages of data" into "likely to be true," for example. When several of them are at work at the same time, it becomes quite difficult to see what's actually going on—a funhouse mirror reflecting a funhouse mirror reflecting reality.

This distorting effect is one of the challenges posed by personalized filters. Like a lens, the filter bubble invisibly transforms the world we experience by controlling what we see and don't see. It interferes with the interplay between our mental processes and our external environment. In some ways, it can act like a magnifying glass, helpfully expanding our view of a niche area of knowledge. But at the same time, personalized filters limit what we are exposed to and therefore affect the way we think and learn. They can upset the delicate cognitive balance that helps us make good decisions and come up with new ideas. And because creativity is also a result of this interplay between mind and environment, they can get in the way of innovation. If we want to know what the world really looks like, we have to understand how filters shape and skew our view of it.

A Fine Balance

It's become a bit in vogue to pick on the human brain. We're "predictably irrational," in the words of behavioral economist Dan Ariely's bestselling book. *Stumbling on Happiness* author Dan Gilbert presents volumes of data to demonstrate that we're terrible at figuring out what makes us happy. Like audience members at a magic show, we're easily conned, manipulated, and misdirected.

All of this is true. But as *Being Wrong* author Kathryn Schulz points out, it's only one part of the story. Human beings may be a walking bundle of miscalculations, contradictions, and irrationalities, but we're built that way for a reason: The same cognitive processes that lead us down the road to error and tragedy are the root of our intelligence and our ability to cope with and survive in a changing world. We pay attention to our mental processes when they fail, but that distracts us from the fact that most of the time, our brains do amazingly well.

The mechanism for this is a cognitive balancing act. Without our ever thinking about it, our brains tread a tightrope between learning too much from the past and incorporating too much new information from the present. The ability to walk this line—to adjust to the demands of different environments and modalities—is one of human cognition's most astonishing traits. Artificial intelligence has yet to come anywhere close.

In two important ways, personalized filters can upset this cognitive balance between strengthening our existing ideas and acquiring new ones. First, the filter bubble surrounds us with ideas with which we're already familiar (and already agree), making us overconfident in our mental frameworks. Second, it removes from our environment some of the key prompts that make us want to learn. To understand how, we have to look at what's being balanced in the first place, starting with how we acquire and store information.

Filtering isn't a new phenomenon. It's been around for millions of years—indeed, it was around before humans even existed. Even for animals with rudimentary senses, nearly all of the information coming in through their senses is meaningless, but a tiny sliver is important and sometimes life-preserving. One of the primary functions of the brain is to identify that sliver and decide what to do about it.

In humans, one of the first steps is to massively compress the data. As Nassim Nicholas Taleb says, "Information wants to be reduced," and every second we reduce a lot of it—compressing most of what our eyes see and ears hear into concepts that capture the gist. Psychologists call these concepts *schemata* (one of them is a *schema*), and they're beginning to be able to identify particular neurons or sets of neurons that correlate with each one—firing, for example, when you recognize a particular object, like a chair. Schemata ensure that we aren't constantly seeing the world anew: Once we've identified something as a chair, we know how to use it.

We don't do this only with objects; we do it with ideas as well. In a study of how people read the news, researcher Doris Graber found that stories were relatively quickly converted into schemata for the purposes of memorization. "Details that do not seem essential at the time and much of the context of a story are routinely pared," she writes in her book *Processing the News*. "Such leveling and sharpening involves condensation of all features of a story." Viewers of a news segment on a child killed by a stray bullet might remember the child's appearance and tragic background, but not the reportage that overall crime rates are down.

Schemata can actually get in the way of our ability to directly observe what's happening. In 1981, researcher Claudia Cohen instructed subjects to watch a video of a woman celebrating her birthday. Some are told that she's a waitress, while others are told she's a librarian. Later, the groups are asked to reconstruct the scene. The people who are told she's a waitress remember her having a beer; those told she was a librarian remember her wearing glasses and listening to classical music (the video shows her doing all three). The information that didn't jibe with her profession was more often forgotten. In some cases, schemata are so powerful they can even lead to information being fabricated: Doris Graber, the news researcher, found that up to a third of her forty-eight subjects had added details to their memories of twelve television news stories shown to them, based on the schemata those stories activated.

Once we've acquired schemata, we're predisposed to strengthen them. Psychological researchers call this confirmation bias—a tendency to believe things that reinforce our existing views, to see what we want to see.

One of the first and best studies of confirmation bias comes from the end of the college football season in 1951—Princeton versus Dartmouth. Princeton hadn't lost a game all season. Its quarterback, Dick Kazmaier, had just been on the cover of *Time*. Things started off pretty rough, but after Kazmaier was sent off the field in the second quarter with a broken nose, the game got really dirty. In the ensuing melee, a Dartmouth player ended up with a broken leg.

Princeton won, but afterward there were recriminations in both college's papers. Princetonians blamed Dartmouth for starting the low blows; Dartmouth thought Princeton had an ax to grind once their quarterback got hurt. Luckily, there were some psychologists on hand to make sense of the conflicting versions of events.

They asked groups of students from both schools who hadn't seen the game to watch a film of it and count how many infractions each side made. Princeton students, on average, saw 9.8 infractions by Dartmouth; Dartmouth students thought their team was guilty of only 4.3. One Dartmouth alumnus who received a copy of the film complained that his version must be missing parts—he didn't see any of the roughhousing he'd heard about. Boosters of each school saw what they wanted to see, not what was actually on the film.

Philip Tetlock, a political scientist, found similar results when he invited a variety of academics and pundits into his office and asked them to make predictions about the future in their areas of expertise. Would the Soviet Union fall in the next ten years? In what year would the U.S. economy start growing again? For ten years, Tetlock kept asking these questions. He asked them not only of experts, but also of folks he'd brought in off the street—plumbers and schoolteachers with no special expertise in politics or history. When he finally compiled the results, even he was surprised. It wasn't just that the normal folks' predictions beat the experts'. The experts' predictions weren't even close.

Why? Experts have a lot invested in the theories they've developed to explain the world. And after a few years of working on them, they tend to see them everywhere. For example, bullish stock analysts banking on rosy financial scenarios were unable to identify the housing bubble that nearly bankrupted the economy—even though the trends that drove it were pretty clear to anyone looking. It's not just that experts are vulnerable to confirmation bias—it's that they're *especially* vulnerable to it.

No schema is an island: Ideas in our heads are connected in networks and hierarchies. *Key* isn't a useful concept without *lock, door,* and a slew of other supporting ideas. If we change these concepts too quickly—altering our concept of *door* without adjusting *lock,* for example—we could end up removing or altering ideas that other ideas depend on and have the whole

system come crashing down. Confirmation bias is a conservative mental force helping to shore up our schemata against
erosion.

Learning, then, is a balance. Jean Piaget, one of the major
figures in developmental psychology, describes it as a process of
assimilation and accommodation. Assimilation happens when
children adapt objects to their existing cognitive structures—as
when an infant identifies every object placed in the crib as
something to suck on. Accommodation happens when we
adjust our schemata to new information—"Ah, this isn't something to suck on, it's something to make a noise with!" We
modify our schemata to fit the world and the world to fit our
schemata, and it's in properly balancing the two processes that
growth occurs and knowledge is built.

The filter bubble tends to dramatically amplify confirmation
bias—in a way, it's designed to. Consuming information that
conforms to our ideas of the world is easy and pleasurable; consuming information that challenges us to think in new ways or
question our assumptions is frustrating and difficult. This is
why partisans of one political stripe tend not to consume the
media of another. As a result, an information environment built
on click signals will favor content that supports our existing
notions about the world over content that challenges them.

During the 2008 presidential campaign, for example, rumors
swirled persistently that Barack Obama, a practicing Christian,
was a follower of Islam. E-mails circulated to millions, offering
"proof" of Obama's "real" religion and reminding voters that
Obama spent time in Indonesia and had the middle name
Hussein. The Obama campaign fought back on television and

encouraged its supporters to set the facts straight. But even a front-page scandal about his Christian priest, Rev. Jeremiah Wright, was unable to puncture the mythology. Fifteen percent of Americans stubbornly held on to the idea that Obama was a Muslim.

That's not so surprising—Americans have never been very well informed about our politicians. What's perplexing is that since the election, the percentage of Americans who hold that belief has nearly doubled, and the increase, according to data collected by the Pew Charitable Trusts, has been greatest among people who are college educated. People with some college education were more likely in some cases to believe the story than people with none—a strange state of affairs.

Why? According to the *New Republic*'s Jon Chait, the answer lies with the media: "Partisans are more likely to consume news sources that confirm their ideological beliefs. People with more education are more likely to follow political news. Therefore, people with more education can actually become mis-educated." And while this phenomenon has always been true, the filter bubble automates it. In the bubble, the proportion of content that validates what you know goes way up.

Which brings us to the second way the filter bubble can get in the way of learning: It can block what researcher Travis Proulx calls "meaning threats," the confusing, unsettling occurrences that fuel our desire to understand and acquire new ideas.

Researchers at the University of California at Santa Barbara asked subjects to read two modified versions of "The Country Doctor," a strange, dreamlike short story by Franz Kafka. "A seriously ill man was waiting for me in a village ten miles distant," begins the story. "A severe snowstorm filled the space between him and me." The doctor has no horse, but when he goes to the stable, it's warm and there's a horsey scent. A belligerent groom hauls himself out of the muck and offers to help the doctor. The groom calls two horses and attempts to rape the doctor's maid, while the doctor is whisked to the patient's house in a snowy instant. And that's just the beginning—the weirdness escalates. The story concludes with a series of non sequiturs and a cryptic aphorism: "Once one responds to a false alarm on the night bell, there's no making it good again—not ever."

The Kafka-inspired version of the story includes meaning threats—incomprehensible events that threaten readers' expectations about the world and shake their confidence in their ability to understand. But the researchers also prepared another version of the story with a much more conventional narrative, complete with a happily-ever-after ending and appropriate, cartoony illustrations. The mysteries and odd occurrences are explained. After reading one version or the other, the study's participants were asked to switch tasks and identify patterns in a set of numbers. The group that read the version adopted from Kafka did nearly twice as well—a dramatic increase in the ability to identify and acquire new patterns. "The key to our study is that our participants were surprised by the series of unexpected events, and they had no way to make sense of them," Proulx wrote. "Hence, they strived to make sense of something else."

For similar reasons, a filtered environment could have consequences for curiosity. According to psychologist George Lowenstein, curiosity is aroused when we're presented with an "information gap." It's a sensation of deprivation: A present's wrapping deprives us of the knowledge of what's in it, and as a result we become curious about its contents. But to feel curiosity, we have to be conscious that something's being hidden. Because the filter bubble hides things invisibly, we're not as compelled to learn about what we don't know.

As University of Virginia media studies professor and Google expert Siva Vaidhyanathan writes in "The Googlization of Everything": "Learning is by definition an encounter with what you don't know, what you haven't thought of, what you couldn't conceive, and what you never understood or entertained as possible. It's an encounter with what's other—even with otherness as such. The kind of filter that Google interposes between an Internet searcher and what a search yields shields the searcher from such radical encounters." Personalization is about building an environment that consists entirely of the adjacent unknown—the sports trivia or political punctuation marks that don't really shake our schemata but *feel* like new information. The personalized environment is very good at answering the questions we have but not at suggesting questions or problems that are out of our sight altogether. It brings to mind the famous Pablo Picasso quotation: "Computers are useless. They can only give you answers."

Stripped of the surprise of unexpected events and associations, a perfectly filtered world would provoke less learning. And there's another mental balance that personalization can upset: the balance between open-mindedness and focus that makes us creative.

The Adderall Society

The drug Adderall is a mixture of amphetamines. Prescribed for attention deficit disorder, it's become a staple for thousands of overscheduled, sleep-deprived students, allowing them to focus for long stretches on a single arcane research paper or complex lab assignment.

For people without ADD, Adderall also has a remarkable effect. On Erowid, an online forum for recreational drug users and "mind hackers," there's post after post of testimonials to the drug's power to extend focus. "The part of my brain that makes me curious about whether I have new e-mails in my inbox apparently shut down," author Josh Foer wrote in an article on *Slate*. "Normally, I can only stare at my computer screen for about 20 minutes at a time. On Adderall, I was able to work in hourlong chunks."

In a world of constant interruptions, as work demands only increase, Adderall is a compelling value proposition. Who couldn't use a little cognitive boost? Among the vocal class of neuroenhancement proponents, Adderall and drugs like it may even be the key to our economic future. "If you're a fifty-five-year-old in Boston, you have to compete with a twenty-six-year-old from Mumbai now, and those kinds of pressures [to use enhancing drugs] are only going to grow," Zack Lynch of the neurotech consulting firm NeuroInsights told a *New Yorker* correspondent.

But Adderall also has some serious side effects. It's addictive. It dramatically boosts blood pressure. And perhaps most important, it seems to decrease associative creativity. After trying Adderall for a week, Foer was impressed with its powers, cranking out pages and pages of text and reading through dense scholarly articles. But, he wrote, "it was like I was thinking with blinders on." "With this drug," an Erowid experimenter wrote, "I become calculating and conservative. In the words of one friend, I think 'inside the box.'" Martha Farah, the director of the University of Pennsylvania's Center for Cognitive Neuroscience, has bigger worries: "I'm a little concerned that we could be raising a generation of very focused accountants."

Like many psychoactive drugs, we still know little about why Adderall has the effects it has—or even entirely what the effects are. But the drug works in part by increasing levels of the neurotransmitter norepinephrine, and norepinephrine has some very particular effects: For one thing, it reduces our sensitivity to new stimuli. ADHD patients call the problem hyperfocus—a trancelike, "zoned out" ability to focus on one thing to the exclusion of everything else.

On the Internet, personalized filters could promote the same kind of intense, narrow focus you get from a drug like Adderall. If you like yoga, you get more information and news about yoga—and less about, say, bird-watching or baseball.

In fact, the search for perfect relevance and the kind of serendipity that promotes creativity push in opposite directions. "If you like this, you'll like that" can be a useful tool, but it's not a source for creative ingenuity. By definition, ingenuity comes from the juxtaposition of ideas that are far apart, and relevance comes from finding ideas that are similar. Personalization, in other words, may be driving us toward an Adderall society, in which hyperfocus displaces general knowledge and synthesis.

Personalization can get in the way of creativity and innovation in three ways. First, the filter bubble artificially limits the size of our "solution horizon"—the mental space in which we search for solutions to problems. Second, the information environment inside the filter bubble will tend to lack some of the key traits that spur creativity. Creativity is a context-dependent trait: We're more likely to come up with new ideas in some environments than in others; the contexts that filtering creates aren't the ones best suited to creative thinking. Finally, the filter bubble encourages a more passive approach to acquiring information, which is at odds with the kind of exploration that leads to discovery. When your doorstep is crowded with salient content, there's little reason to travel any farther.

In his seminal book *The Act of Creation*, Arthur Koestler describes creativity as "bisociation"—the intersection of two "matrices" of thought: "Discovery is an analogy no one has ever seen before." Friedrich Kekule's epiphany about the structure of a benzene molecule after a daydream about a snake eating its tail is an example. So is Larry Page's insight to apply the technique of academic citation to search. "Discovery often means simply the uncovering of something which has always been there but was hidden from the eye by the blinkers of habit," Koestler wrote. Creativity "uncovers, selects, re-shuffles, combines, synthesizes already existing facts, ideas, faculties, (and) skills."

While we still have little insight into exactly where different words, ideas, and associations are located physically in the brain, researchers are beginning to be able to map the terrain abstractly. They know that when you feel as though a word is on the tip of your tongue, it usually is. And they can tell that some concepts are much further apart than others, in neural connections if not in actual physical brain space. Researcher Hans Eysenck has found evidence that the individual differences in how people do this mapping—how they connect concepts together—are the key to creative thought.

In Eysenck's model, creativity is a search for the right set of ideas to combine. At the center of the mental search space are the concepts most directly related to the problem at hand, and as you move outward, you reach ideas that are more tangentially connected. The solution horizon delimits where we stop searching. When we're instructed to "think outside the box," the box represents the solution horizon, the limit of the conceptual area that we're operating in. (Of course, solution horizons that are too wide are a problem, too, because more ideas means exponentially more combinations.)

Programmers building artificially intelligent chess masters learned the importance of the solution horizon the hard way. The early ones trained the computer to look at every possible combination of moves. This resulted in an explosion of possibilities, which in turn meant that even very powerful computers could only look a limited number of moves ahead. Only when programmers discovered heuristics that allowed the computers to discard some of the moves did they become powerful enough to win against the grand masters of chess. Narrowing the solution horizon, in other words, was key.

In a way, the filter bubble is a prosthetic solution horizon: It provides you with an information environment that's highly relevant to whatever problem you're working on. Often, this'll be highly useful: When you search for "restaurant," it's likely that you're also interested in near synonyms like "bistro" or "café." But when the problem you're solving requires the bisociation of ideas that are indirectly related—as when Page applied the logic of academic citation to the problem of Web search—the filter bubble may narrow your vision too much.

What's more, some of the most important creative breakthroughs are spurred by the introduction of the entirely random ideas that filters are designed to rule out.

The word *serendipity* originates with the fairy tale "The Three Princes of Serendip," who are continually setting out in search of one thing and finding another. In what researchers call the evolutionary view of innovation, this element of random chance isn't just fortuitous, it's necessary. Innovation requires serendipity.

Since the 1960s, a group of researchers, including Donald Campbell and Dean Simonton, has been pursuing the idea that at a cultural level the process of developing new ideas looks a lot like the process of developing new species. The evolutionary process can be summed up in four words: "Blind variation, selective retention." Blind variation is the process by which mutations and accidents change genetic code, and it's blind because it's chaotic—it's variation that doesn't know where it's going. There's no intent behind it, nowhere in particular that it's headed—it's just the random recombination of genes. Selective retention is the process by which some of the results of blind variation—the offspring—are "retained" while others perish.

When problems become acute enough for enough people, the argument goes, the random recombination of ideas in millions of heads will tend to produce a solution. In fact, it'll tend to produce the same solution in multiple different heads around the same time.

The way we selectively combine ideas isn't always blind: As Eysenck's "solution horizon" suggests, we don't try to solve our problems by combining every single idea with every other idea in our heads. But when it comes to really new ideas, innovation is in fact often blind. Aharon Kantorovich and Yuval Ne'eman are two historians of science whose work focuses on paradigm shifts, like the move from Newtonian to Einsteinian physics. They argue that "normal science"—the day-to-day process of experimentation and prediction—doesn't benefit much from blind variation, because scientists tend to discard random combinations and strange data.

But in moments of major change, when our whole way of looking at the world shifts and recalibrates, serendipity is often at work. "Blind discovery is a necessary condition for scientific revolution," they write, for a simple reason: The Einsteins and Copernicuses and Pasteurs of the world often have no idea what they're looking for. The biggest breakthroughs are sometimes the ones that we least expect.

The filter bubble still offers the opportunity for some serendipity, of course. If you're interested in football and local politics, you might still see a story about a play that gives you an idea about how to win the mayoral campaign. But overall, there will tend to be fewer random ideas around—that's part of the point. For a quantified system like a personal filter, it's nearly impossible to sort the usefully serendipitous and randomly provocative from the just plain irrelevant.

The second way in which the filter bubble can dampen creativity is by removing some of the diversity that prompts us to think in new and innovative ways. In one of the standard creativity tests developed by Karl Duncker in 1945, a researcher hands a subject a box of thumbtacks, a candle, and a book of matches. The subject's job is to attach the candle to the wall so that, when lit, it doesn't drip on the table below (or set the wall on fire). Typically, people try to tack the candle to the wall, or glue it by melting it, or by building complex structures out of the wall with wax and tacks. But the solution (spoiler alert!) is quite simple: Tack the inside of the box to the wall, then place the candle in the box.

Duncker's test gets at one of the key impediments to creativity, what early creativity researcher George Katona described as the reluctance to "break perceptual set." When you're handed a box full of tacks, you'll tend to register the box itself as a container. It takes a conceptual leap to see it as a platform, but even a small change in the test makes that much more likely: If subjects receive the box separately from the tacks, they tend to see the solution much more quickly.

The process of mapping "thing with tacks in it" to the schema "container" is called coding; creative candle-platform-builders are those who are able to code objects and ideas in multiple ways. Coding, of course, is very useful: It tells you what you can do with the object; once you've decided that something fits in the "chair" schema, you don't have to think twice about sitting on it. But when the coding is too narrow, it impedes creativity.

In study after study, creative people tend to see things in many different ways and put them in what researcher Arthur Cropley calls "wide categories." The notes from a 1974 study in which participants were told to make groups of similar objects offers an amusing example of the trait in excess: "Subject 30, a writer, sorted a total of 40 objects. . . . In response to the candy cigar, he sorted the pipe, matches, cigar, apple, and sugar cubes, explaining that all were related to consumption. In response to the apple, he sorted only the wood block with the nail driven into it, explaining that the apple represented health and vitality (or yin) and that the wood block represented a coffin with a nail in it, or death (or yang). Other sortings were similar."

It's not just artists and writers who use wide categories. As Cropley points out in *Creativity in Education and Learning*, the physicist Niels Bohr famously demonstrated this type of creative dexterity when he was given a university exam at the University of Copenhagen in 1905. One of the questions asked students to explain how they would use a barometer (an instrument that measures atmospheric pressure) to measure the height of a building. Bohr clearly knew what the instructor was going for: Students were supposed to check the atmospheric pressure at the top and bottom of the building and do some math. Instead, he suggested a more original method: One could tie a string to the barometer, lower it, and measure the string—thinking of the instrument as a "thing with weight."

The unamused instructor gave him a failing grade—his answer, after all, didn't show much understanding of physics. Bohr appealed, this time offering four solutions: You could throw the barometer off the building and count the seconds until it hit the ground (barometer as mass); you could measure the length of the barometer and of its shadow, then measure the building's shadow and calculate its height (barometer as object with length); you could tie the barometer to a string and swing it at ground level and from the top of the building to determine the difference in gravity (barometer as mass again); or you could use it to calculate air pressure. Bohr finally passed, and one moral of the story is pretty clear: Avoid smartass physicists. But the episode also explains why Bohr was such a brilliant innovator: His ability to see objects and concepts in many different ways made it easier for him to use them to solve problems.

The kind of categorical openness that supports creativity also correlates with certain kinds of luck. While science has yet to find that there are people whom the universe favors—ask people to guess a random number, and we're all about equally bad at it—there are some traits that people who consider themselves to be lucky share. They're more open to new experiences and new people. They're also more distractable.

Richard Wiseman, a luck researcher at the University of Hertfordshire in England, asked groups of people who considered themselves to be lucky and unlucky to flip through a doctored newspaper and count the number of photographs in it. On the second page, a big headline said, "Stop counting—there are 43 pictures." Another page offered 150 British pounds to readers who noticed it. Wiseman described the results: "For the most part, the unlucky would just flip past these things. Lucky people would flip through and laugh and say, 'There are 43 photos. That's what it says. Do you want me to bother counting?' We'd say, 'Yeah, carry on.' They'd flip some more and say, 'Do I get my 150 pounds?' Most of the unlucky people didn't notice."

As it turns out, being around people and ideas unlike oneself is one of the best ways to cultivate this sense of open-mindedness and wide categories. Psychologists Charlan Nemeth and Julianne Kwan discovered that bilinguists are more creative than monolinguists—perhaps because they have to get used to the proposition that things can be viewed in several different ways. Even forty-five minutes of exposure to a different culture can boost creativity: When a group of American students was shown a slideshow about China as opposed to one about the United States, their scores on several creativity tests went up. In companies, the people who interface with multiple different units tend to be greater sources of innovation than people who interface only with their own. While nobody knows for certain what causes this effect, it's likely that foreign ideas help us break open our categories.

But the filter bubble isn't tuned for a diversity of ideas or of people. It's not designed to introduce us to new cultures. As a result, living inside it, we may miss some of the mental flexibility and openness that contact with difference creates.

But perhaps the biggest problem is that the personalized Web encourages us to spend less time in discovery mode in the first place.

The Age of Discovery

In *Where Good Ideas Come From*, science author Steven Johnson offers a "natural history of innovation," in which he inventories and elegantly illustrates how creativity arises. Creative environments often rely on "liquid networks" where different ideas can collide in different configurations. They arrive through serendipity—we set out looking for the answer to one problem and find another—and as a result, ideas emerge frequently in places where random collision is more likely to occur. "Innovative environments," he writes, "are better at helping their inhabitants explore the adjacent possible"—the bisociated area in which existing ideas combine to produce new ones— "because they expose a wide and diverse sample of spare parts— mechanical or conceptual—and they encourage novel ways of recombining those parts."

His book is filled with examples of these environments, from primordial soup to coral reefs and high-tech offices, but Johnson continually returns to two: the city and the Web.

"For complicated historical reasons," he writes, "they are both environments that are powerfully suited for the creation, diffusion, and adoption of good ideas."

There's no question that Johnson *was* right: The old, unpersonalized web offered an environment of unparalleled richness and diversity. "Visit the 'serendipity' article in Wikipedia," he writes, and "you are one click away from entries on LSD, Teflon, Parkinson's disease, Sri Lanka, Isaac Newton, and about two hundred other topics of comparable diversity."

But the filter bubble has dramatically changed the informational physics that determines which ideas we come in contact with. And the new, personalized Web may no longer be as well suited for creative discovery as it once was.

In the early days of the World Wide Web, when Yahoo was its king, the online terrain felt like an unmapped continent, and its users considered themselves discoverers and explorers. Yahoo was the village tavern where sailors would gather to swap tales about what strange beasts and distant lands they found out at sea. "The shift from exploration and discovery to the intent-based search of today was inconceivable," an early Yahoo editor told search journalist John Battelle. "Now, we go online expecting everything we want to find will be there. That's a major shift."

This shift from a discovery-oriented Web to a search and retrieval–focused Web mirrors one other piece of the research surrounding creativity. Creativity experts mostly agree that it's a process with at least two key parts: Producing novelty requires a lot of divergent, generative thinking—the reshuffling and recombining that Koestler describes. Then there's a winnowing process—convergent thinking—as we survey the options for one that'll fit the situation. The serendipitous Web attributes that Johnson praises—the way one can hop from article to article on Wikipedia—are friendly to the divergent part of that process.

But the rise of the filter bubble means that increasingly the convergent, synthetic part of the process is built in. Battelle calls Google a "database of intentions," each query representing something that someone wants to do or know or buy. Google's core mission, in many ways, is to transform those intentions into actions. But the better it gets at that, the worse it'll be at providing serendipity, which, after all, is the process of stumbling across the *un*intended. Google is great at helping us find what we know we want, but not at finding what we *don't* know we want.

To some degree, the sheer volume of information available mitigates this effect. There's far more online content to choose from than there was in even the largest libraries. For an enterprising informational explorer, there's endless terrain to cover. But one of the prices of personalization is that we become a bit more passive in the process. The better it works, the less exploring we have to do.

David Gelernter, a Yale professor and early supercomputing visionary, believes that computers will only serve us well when they can incorporate dream logic. "One of the hardest, most fascinating problems of this cyber-century is how to add 'drift' to the net," he writes, "so that your view sometimes wanders (as your mind wanders when you're tired) into places you hadn't planned to go. Touching the machine brings the original topic back. We need help overcoming rationality sometimes, and allowing our thoughts to wander and metamorphose as they do in sleep." To be truly helpful, algorithms may need to work more like the fuzzy-minded, nonlinear humans they're supposed to serve.

On California Island

In 1510, the Spanish writer Garci Rodriguez de Montalvo published a swashbuckling *Odyssey*-like novel, *The Exploits of*

Esplandian, which included a description of a vast island called California:

> On the right hand from the Indies exists an island called California very close to a side of the Earthly Paradise; and it was populated by black women, without any man existing there, because they lived in the way of the Amazons. They had beautiful and robust bodies, and were brave and very strong. Their island was the strongest of the World, with its cliffs and rocky shores. Their weapons were golden and so were the harnesses of the wild beasts that they were accustomed to domesticate and ride, because there was no other metal in the island than gold.

Rumors of gold propelled the legend of the island of California across Europe, prompting adventurers throughout the continent to set off in search of it. Hernán Cortés, the Spanish conquistador who led the colonization of the Americas, requested money from Spain's king to lead a worldwide hunt. And when he landed in what we now know as Baja California in 1536, he was certain he'd found the place. It wasn't until one of his navigators, Francisco de Ulloa, traveled up the Gulf of California to the mouth of the Colorado river that it became clear to Cortez that, gold or no, he hadn't found the mythical island.

Despite this discovery, however, the idea that California was an island persisted for several more centuries. Other explorers discovered Puget Sound, near Vancouver, and were certain that it must connect to Baja. Dutch maps from the 1600s routinely show a distended long fragment off the coast of America stretching half the length of the continent. It took Jesuit missionaries literally marching inland and never reaching the other side to fully repudiate the myth.

It may have persisted for one simple reason: There was no sign on the maps for "don't know," and so the distinction between geographic guesswork and sights that had been witnessed firsthand became blurred. One of history's major cartographic errors, the island of California reminds us that it's not what we don't know that hurts us as much as what we don't know we don't know—what ex–secretary of defense Donald Rumsfeld famously called the unknown unknowns.

This is one other way that personalized filters can interfere with our ability to properly understand the world: They alter our sense of the map. More unsettling, they often remove its blank spots, transforming known unknowns into unknown ones.

Traditional, unpersonalized media often offer the promise of representativeness. A newspaper editor isn't doing his or her job properly unless to some degree the paper is representative of the news of the day. This is one of the ways one can convert an unknown unknown into a known unknown. If you leaf through the paper, dipping into some articles and skipping over most of them, you at least know there are stories, perhaps whole sections, that you passed over. Even if you don't read the article, you notice the headline about a flood in Pakistan—or maybe you're just reminded that, yes, there is a Pakistan.

In the filter bubble, things look different. You don't see the things that don't interest you at all. You're not even latently aware that there are major events and ideas you're missing. Nor can you take the links you do see and assess how representative they are without an understanding of what the broader environment from which they were selected looks like. As any statistician will tell you, you can't tell how biased the sample is from looking at the sample alone: You need something to compare it to.

As a last resort, you might look at your selection and ask yourself if it *looks* like a representative sample. Are there conflicting views? Are there different takes, and different kinds of people reflecting? Even this is a blind alley, however, because with an information set the size of the Internet, you get a kind of fractal diversity: at any level, even within a very narrow information spectrum (atheist goth bowlers, say) there are lots of voices and lots of different takes.

We're never able to experience the whole world at once. But the best information tools give us a sense of where we stand in it—literally, in the case of a library, and figuratively in the case of a newspaper front page. This was one of the CIA's primary errors with Yuri Nosenko. The agency had collected a specialized subset of information about Nosenko without realizing how specialized it was, and thus despite the many brilliant analysts working for years on the case, it missed what would have been obvious from a whole picture of the man.

Because personalized filters usually have no Zoom Out function, it's easy to lose your bearings, to believe the world is a narrow island when in fact it's an immense, varied continent.

The You Loop

I believe this is the quest for what a personal computer really is. It is to capture one's entire life.

—*Gordon Bell*

You have one identity," Facebook founder Mark Zuckerberg told journalist David Kirkpatrick for his book *The Facebook Effect*. "The days of you having a different image for your work friends or coworkers and for the other people you know are probably coming to an end pretty quickly. . . . Having two identities for yourself is an example of a lack of integrity."

A year later, soon after the book had been published, twenty-six-year-old Zuckerberg sat onstage with Kirkpatrick and NPR interviewer Guy Raz at the Computer History Museum in Mountain View, California. "In David's book," Raz said, "you say that people should have one identity. . . . But I behave a different way around my family than I do around my colleagues."

Zuckerberg shrugged. "No, I think that was just a sentence I said."

Raz continued: "Are you the same person right now as when you're with your friends?"

"Uh, yeah," Zuckerberg said. "Same awkward self."

If Mark Zuckerberg were a standard mid-twenty-something, this tangle of views might be par for the course: Most of us don't spend too much time musing philosophically about the nature of identity. But Zuckerberg controls the world's most powerful and widely used technology for managing and expressing who we are. And his views on the matter are central to his vision for the company and for the Internet.

Speaking at an event during New York's Ad Week, Facebook COO Sheryl Sandberg said she expected the Internet to change quickly. "People don't want something targeted to the whole world—they want something that reflects what they want to see and know," she said, suggesting that in three to five years that would be the norm. Facebook's goal is to be at the center of that process—the singular platform through which every other service and Web site incorporates your personal and social data. You have one identity, it's your Facebook identity, and it colors your experience everywhere you go.

It's hard to imagine a more dramatic departure from the early days of the Internet, in which *not* exposing your identity was part of the appeal. In chat rooms and online forums, your gender, race, age, and location were whatever you said they were, and the denizens of these spaces exulted about the way the medium allowed you to shed your skin. Electronic Frontier Foundation (EFF) founder John Perry Barlow dreamed of "creating a world that all may enter without privilege or prejudice accorded by race, economic power, military force, or station of birth." The freedom that this offered anyone who was interested to transgress and explore, to try on different personas for size, felt revolutionary.

As law and commerce have caught up with technology, however, the space for anonymity online is shrinking. You can't hold an anonymous person responsible for his or her actions: Anonymous customers commit fraud, anonymous commenters start flame wars, and anonymous hackers cause trouble. To establish the trust that community and capitalism are built on, you need to know whom you're dealing with.

As a result, there are dozens of companies working on de-anonymizing the Web. PeekYou, a firm founded by the creator of RateMyProfessors.com, is patenting ways of connecting online activities done under a pseudonym with the real name of the person involved. Another company, Phorm, helps Internet service providers use a method called "deep packet inspection" to analyze the traffic that flows through their servers; Phorm aims to build nearly comprehensive profiles of each customer to use for advertising and personalized services. And if ISPs are leery, BlueCava is compiling a database of every computer, smartphone, and online-enabled gadget in the world, which can be tied to the individual people who use them. Even if you're using the highest privacy settings in your Web browser, in other words, your hardware may soon give you away.

These technological developments pave the way for a more persistent kind of personalization than anything we've experienced to date. It also means that we'll increasingly be forced to trust the companies at the center of this process to properly express and synthesize who we really are. When you meet someone in a bar or a park, you look at how they behave and act and form an impression accordingly. Facebook and the other identity services aim to mediate that process online; if they don't do it right, things can get fuzzy and distorted. To personalize well, you have to have the right idea of what represents a person.

There's another tension in the interplay of identity and personalization. Most personalized filters are based on a three-step model. First, you figure out who people are and what they like. Then, you provide them with content and services that best fit them. Finally, you tune to get the fit just right. Your identity shapes your media. There's just one flaw in this logic: Media also shape identity. And as a result, these services may end up creating a good fit between you and your media by changing . . . you. If a self-fulfilling prophecy is a false definition of the world that through one's actions becomes true, we're now on the verge of self-fulfilling identities, in which the Internet's distorted picture of us becomes who we really are.

Personalized filtering can even affect your ability to choose your own destiny. In "Of Sirens and Amish Children," a much-cited tract, information law theorist Yochai Benkler describes how more-diverse information sources make us freer. Autonomy, Benkler points out, is a tricky concept: To be free, you have to be able not only to do what you want, but to know what's possible to do. The Amish children in the title are plaintiffs in a famous court case, *Wisconsin v. Yoder*, whose parents sought to prevent them from attending public school so that they wouldn't be exposed to modern life. Benkler argues that this is a real threat to the children's freedom: Not knowing that it's possible to be an astronaut is just as much a prohibition against becoming one as knowing and being barred from doing so.

Of course, too many options are just as problematic as too few—you can find yourself overwhelmed by the number of options or paralyzed by the paradox of choice. But the basic point remains: The filter bubble doesn't just reflect your identity. It also illustrates what choices you have. Students who go to Ivy League colleges see targeted advertisements for jobs that students at state schools are never even aware of. The personal feeds of professional scientists might feature articles about contests that amateurs never become aware of. By illustrating some possibilities and blocking out others, the filter bubble has a hand in your decisions. And in turn, it shapes who you become.

A Bad Theory of You

The way that personalization shapes identity is still becoming clear—especially because most of us still spend more time consuming broadcast media than personalized content streams. But by looking at how the major filterers think about identity, it's becoming possible to predict what these changes might look like. Personalization requires a theory of what makes a person—of what bits of data are most important to determine who someone is—and the major players on the Web have quite different ways of approaching the problem.

Google's filtering systems, for example, rely heavily on Web history and what you click on (click signals) to infer what you like and dislike. These clicks often happen in an entirely private context: The assumption is that searches for "intestinal gas" and celebrity gossip Web sites are between you and your browser. You might behave differently if you thought other people were going to see your searches. But it's that behavior that determines what content you see in Google News, what ads Google displays—what determines, in other words, Google's theory of you.

The basis for Facebook's personalization is entirely different. While Facebook undoubtedly tracks clicks, its primary way of thinking about your identity is to look at what you share and with whom you interact. That's a whole different kettle of data from Google's: There are plenty of prurient, vain, and embarrassing things we click on that we'd be reluctant to share with all of our friends in a status update. And the reverse is true, too. I'll cop to sometimes sharing links I've barely read—the long investigative piece on the reconstruction of Haiti, the bold political headline—because I like the way it makes me appear to others. The Google self and the Facebook self, in other words, are pretty different people. There's a big difference between "you are what you click" and "you are what you share."

Both ways of thinking have their benefits and drawbacks. With Google's click-based self, the gay teenager who hasn't come out to his parents can still get a personalized Google News feed with pieces from the broader gay community that affirm that he's not alone. But by the same token, a self built on clicks will tend to draw us even more toward the items we're predisposed to look at already—toward our most Pavlovian selves. Your perusal of an article on TMZ.com is filed away, and the next time you're looking at the news, Brad Pitt's marriage drama is more likely to flash on to the screen. (If Google didn't persistently downplay porn, the problem would presumably be far worse.)

Facebook's share-based self is more aspirational: Facebook takes you more at your word, presenting you as you'd like to be seen by others. Your Facebook self is more of a performance, less of a behaviorist black box, and ultimately it may be more prosocial than the bundle of signals Google tracks. But the Facebook approach has its downsides as well—to the extent that Facebook draws on the more public self, it necessarily has less room for private interests and concerns. The same closeted gay teenager's information environment on Facebook might diverge more from his real self. The Facebook portrait remains incomplete.

Both are pretty poor representations of who we are, in part because there is no one set of data that describes who we are. "Information about our property, our professions, our purchases, our finances, and our medical history does not tell the whole story," writes privacy expert Daniel Solove. "We are more than the bits of data we give off as we go about our lives."

Digital animators and robotics engineers frequently run into a problem known as the *uncanny valley*. The uncanny valley is the place where something is lifelike but not convincingly alive, and it gives people the creeps. It's part of why digital animation of real people still hasn't hit the big screens: When an image looks almost like a real person, but not quite, it's unsettling on a basic psychological level. We're now in the uncanny valley of personalization. The doppelgänger selves reflected in our media are a lot like, but not exactly, ourselves. And as we'll see, there are some important things that are lost in the gap between the data and reality.

To start with, Zuckerberg's statement that we have "one identity" simply isn't true. Psychologists have a name for this fallacy: fundamental attribution error. We tend to attribute peoples' behavior to their inner traits and personality rather than to the situations they're placed in. Even in situations where the context clearly plays a major role, we find it hard to separate how someone behaves from who she is.

And to a striking degree, our characteristics are fluid. Someone who's aggressive at work may be a doormat at home. Someone who's gregarious when happy may be introverted when stressed. Even some of our closest-held traits—our disinclination to do harm, for example—can be shaped by context. Groundbreaking psychologist Stanley Milgram demonstrated this when, in an oft-cited experiment at Yale in the 1960s, he got decent ordinary people to apparently electrocute other subjects when given the nod by a man in a lab coat.

There is a reason that we act this way: The personality traits that serve us well when we're at dinner with our family might get in the way when we're in a dispute with a passenger on the train or trying to finish a report at work. The plasticity of the self allows for social situations that would be impossible or intolerable if we always behaved exactly the same way. Advertisers have understood this phenomenon for a long time. In the jargon, it's called *day-parting*, and it's the reason that you don't hear many beer ads as you're driving to work in the morning. People have different needs and aspirations at eight A.M. than they do at eight P.M. By the same token, billboards in the night-life district promote different products than billboards in the residential neighborhoods the same partiers go home to.

On his own Facebook page, Zuckerberg lists "transparency" as one of his top Likes. But there's a downside to perfect transparency: One of the most important uses of privacy is to manage and maintain the separations and distinctions among our different selves. With only one identity, you lose the nuances that make for a good personalized fit.

Personalization doesn't capture the balance between your work self and your play self, and it can also mess with the tension between your aspirational and your current self. How we behave is a balancing act between our future and present selves. In the future, we want to be fit, but in the present, we want the candy bar. In the future, we want to be a well-rounded, well-informed intellectual virtuoso, but right now we want to watch *Jersey Shore*. Behavioral economists call this present bias—the gap between your preferences for your future self and your preferences in the current moment.

The phenomenon explains why there are so many movies in your Netflix queue. When researchers at Harvard and the Analyst Institute looked at people's movie-rental patterns, they were able to watch as people's future aspirations played against their current desires. "Should" movies like *An Inconvenient Truth* or *Schindler's List* were often added to the queue, but there they languished while watchers gobbled up "want" movies like *Sleepless in Seattle*. And when they had to choose three movies to watch instantly, they were less likely to choose "should" movies at all. Apparently there are some movies we'd always rather watch tomorrow.

At its best, media help mitigate present bias, mixing "should" stories with "want" stories and encouraging us to dig into the difficult but rewarding work of understanding complex problems.

But the filter bubble tends to do the opposite: Because it's our present self that's doing all the clicking, the set of preferences it reflects is necessarily more "want" than "should."

The one-identity problem isn't a fundamental flaw. It's more of a bug: Because Zuckerberg thinks you have one identity and you don't, Facebook will do a worse job of personalizing your information environment. As John Battelle told me, "We're so far away from the nuances of what it means to be human, as reflected in the nuances of the technology." Given enough data and enough programmers, the context problem is solvable—and according to personalization engineer Jonathan McPhie, Google is working on it. We've seen the pendulum swing from the anonymity of the early Internet to the one-identity view currently in vogue; the future may look like something in between.

But the one-identity problem illustrates one of the dangers of turning over your most personal details to companies who have a skewed view of what identity is. Maintaining separate identity zones is a ritual that helps us deal with the demands of different roles and communities. And something's lost when, at the end of the day, everything inside your filter bubble looks roughly the same. Your bacchanalian self comes knocking at work; your work anxieties plague you on a night out.

And when we're aware that everything we do enters a permanent, pervasive online record, another problem emerges: The knowledge that what we do affects what we see and how companies see us can create a chilling effect. Genetic privacy expert Mark Rothstein describes how lax regulations around genetic data can actually reduce the number of people willing to be tested for certain diseases: If you might be discriminated against or denied insurance for having a gene linked to Parkinson's disease, it's not unreasonable just to skip the test and the "toxic knowledge" that might result.

In the same way, when our online actions are tallied and added to a record that companies use to make decisions, we might decide to be more cautious in our surfing. If we knew (or even suspected, for that matter) that purchasers of *101 Ways to Fix Your Credit Score* tend to get offered lower-premium credit cards, we'd avoid buying the book. "If we thought that our every word and deed were public," writes law professor Charles Fried, "fear of disapproval or more tangible retaliation might keep us from doing or saying things which we would do or say could we be sure of keeping them to ourselves." As Google expert Siva Vaidhyanathan points out, "F. Scott Fitzgerald's enigmatic Jay Gatsby could not exist today. The digital ghost of Jay Gatz would follow him everywhere."

In theory, the one-identity, context-blind problem isn't impossible to fix. Personalizers will undoubtedly get better at sensing context. They might even be able to better balance long-term and short-term interests. But when they do—when they are able to accurately gauge the workings of your psyche—things get even weirder.

Targeting Your Weak Spots

The logic of the filter bubble today is still fairly rudimentary: People who bought the *Iron Man* DVD are likely to buy *Iron Man II*; people who enjoy cookbooks will probably be interested in cookware. But for Dean Eckles, a doctoral student at Stanford and an adviser to Facebook, these simple recommendations are just the beginning. Eckles is interested in means, not ends: He cares less about what types of products you like than which kinds of arguments might cause you to choose one over another.

Eckles noticed that when buying products—say, a digital camera—different people respond to different pitches. Some people feel comforted by the fact that an expert or product review site will vouch for the camera. Others prefer to go with the product that's most popular, or a money-saving deal, or a brand that they know and trust. Some people prefer what Eckles calls "high cognition" arguments—smart, subtle points that require some thinking to get. Others respond better to being hit over the head with a simple message.

And while most of us have preferred styles of argument and validation, there are also types of arguments that really turn us off. Some people rush for a deal; others think that the deal means the merchandise is subpar. Just by eliminating the persuasion styles that rub people the wrong way, Eckles found he could increase the effectiveness of marketing materials by 30 to 40 percent.

While it's hard to "jump categories" in products—what clothing you prefer is only slightly related to what books you enjoy— "persuasion profiling" suggests that the kinds of arguments you respond to are highly transferrable from one domain to another. A person who responds to a "get 20% off if you buy NOW" deal for a trip to Bermuda is much more likely than someone who doesn't to respond to a similar deal for, say, a new laptop.

If Eckles is right—and research so far appears to be validating his theory—your "persuasion profile" would have a pretty significant financial value. It's one thing to know how to pitch products to you in a specific domain; it's another to be able to improve the hit rate anywhere you go. And once a company like Amazon has figured out your profile by offering you different kinds of deals over time and seeing which ones you responded to, there's no reason it couldn't then sell that information to other companies. (The field is so new that it's not clear if there's a correlation between persuasion styles and demographic traits, but obviously that could be a shortcut as well.)

There's plenty of good that could emerge from persuasion profiling, Eckles believes. He points to DirectLife, a wearable coaching device by Philips that figures out which arguments get people eating more healthily and exercising more regularly. But he told me he's troubled by some of the possibilities. Knowing what kinds of appeals specific people respond to gives you power to manipulate them on an individual basis.

With new methods of "sentiment analysis, it's now possible to guess what mood someone is in. People use substantially more positive words when they're feeling up; by analyzing enough of your text messages, Facebook posts, and e-mails, it's possible to tell good days from bad ones, sober messages from drunk ones (lots of typos, for a start). At best, this can be used to provide content that's suited to your mood: On an awful day in the near future, Pandora might know to preload *Pretty Hate Machine* for you when you arrive. But it can also be used to take advantage of your psychology.

Consider the implications, for example, of knowing that particular customers compulsively buy things when stressed or when they're feeling bad about themselves, or even when they're a bit tipsy. If persuasion profiling makes it possible for a coaching device to shout "you can do it" to people who like positive reinforcement, in theory it could also enable politicians to make appeals based on each voter's targeted fears and weak spots.

Infomercials aren't shown in the middle of the night only because airtime then is cheap. In the wee hours, most people are especially suggestible. They'll spring for the slicer-dicer that they'd never purchase in the light of day. But the three A.M. rule is a rough one—presumably, there are times in all of our daily lives when we're especially inclined to purchase whatever's put in front of us. The same data that provides personalized content can be used to allow marketers to find and manipulate your personal weak spots. And this isn't a hypothetical possibility: Privacy researcher Pam Dixon discovered that a data company called PK List Management offers a list of customers titled "Free to Me—Impulse Buyers"; those listed are described as being highly susceptible to pitches framed as sweepstakes.

If personalized persuasion works for products, it can also work for ideas. There are undoubtedly times and places and styles of argument that make us more susceptible to believe what we're told. Subliminal messaging is illegal because we recognize there are some ways of making an argument that are essentially cheating; priming people with subconsciously flashed words to sell them things isn't a fair game. But it's not such a stretch to imagine political campaigns targeting voters at times when they can circumvent our more reasonable impulses.

We intuitively understand the power in revealing our deep motivations and desires and how we work, which is why most of us only do that in day-to-day life with people whom we really trust. There's a symmetry to it: You know your friends about as well as they know you. Persuasion profiling, on the other hand, can be done invisibly—you need not have any knowledge that this data is being collected from you—and therefore it's asymmetrical. And unlike some forms of profiling that take place in plain sight (like Netflix), persuasion profiling is handicapped when it's revealed. It's just not the same to hear an automated coach say "You're doing a great job! I'm telling you that because you respond well to encouragement!"

So you don't necessarily see the persuasion profile being made. You don't see it being used to influence your behavior. And the companies we're turning over this data to have no legal obligation to keep it to themselves. In the wrong hands, persuasion profiling gives companies the ability to circumvent your rational decision making, tap into your psychology, and draw out your compulsions. Understand someone's identity, and you're better equipped to influence what he or she does.

A Deep and Narrow Path

Someday soon, Google Vice President Marissa Mayer says, the company hopes to make the search box obsolete. "The next step of search is doing this automatically," Eric Schmidt said in 2010. "When I walk down the street, I want my smartphone to be doing searches constantly—'did you know?' 'did you know?' 'did you know?' 'did you know?'" In other words, your phone should figure out what you would like to be searching for before you do.

In the fast-approaching age of search without search, identity drives media. But the personalizers haven't fully grappled with a parallel fact: Media also shapes identity. Political scientist Shanto Iyengar calls one of primary factors accessibility bias, and in a paper titled "Experimental Demonstrations of the 'Not-So-Minimal' Consequences of Television News,'" in 1982, he demonstrated how powerful the bias is. Over six days, Iyengar asked groups of New Haven residents to watch episodes of a TV news program, which he had doctored to include different segments for each group.

Afterward, Iyengar asked subjects to rank how important issues like pollution, inflation, and defense were to them. The shifts from the surveys they'd filled out before the study were dramatic: "Participants exposed to a steady stream of news about defense or about pollution came to believe that defense or pollution were more consequential problems," Iyengar wrote. Among the group that saw the clips on pollution, the issue moved from fifth out of six in priority to second.

Drew Westen, a neuropsychologist whose focus is on political persuasion, demonstrates the strength of this priming effect by asking a group of people to memorize a list of words that include *moon* and *ocean*. A few minutes later, he changes topics and asks the group which detergent they prefer. Though he hasn't mentioned the word, the group's show of hands indicates a strong preference for Tide.

Priming isn't the only way media shape our identities. We're also more inclined to believe what we've heard before. In a

1977 study by Hasher and Goldstein, participants were asked to read sixty statements and mark whether they were true or false. All of the statements were plausible, but some of them ("French horn players get cash bonuses to stay in the Army") were true; others ("Divorce is only found in technically advanced societies") weren't. Two weeks later, they returned and rated a second batch of statements in which some of the items from the first list had been repeated. By the third time, two weeks after that, the subjects were far more likely to believe the repeated statements. With information as with food, we are what we consume.

All of these are basic psychological mechanisms. But combine them with personalized media, and troubling things start to happen. Your identity shapes your media, and your media then shapes what you believe and what you care about. You click on a link, which signals an interest in something, which means you're more likely to see articles about that topic in the future, which in turn prime the topic for you. You become trapped in a you loop, and if your identity is misrepresented, strange patterns begin to emerge, like reverb from an amplifier.

If you're a Facebook user, you've probably run into this problem. You look up your old college girlfriend Sally, mildly curious to see what she is up to after all these years. Facebook interprets this as a sign that you're interested in Sally, and all of a sudden her life is all over your news feed. You're still mildly curious, so you click through on the new photos she's posted of her kids and husband and pets, confirming Facebook's hunch. From Facebook's perspective, it looks as though you have a relationship with this person, even if you haven't communicated in years. For months afterward, Sally's life is far more prominent than your actual relationship would indicate. She's a "local maximum": Though there are people whose posts you're far more interested in, it's her posts that you see.

In part, this feedback effect is due to what early Facebook employee and venture capitalist Matt Cohler calls the local-maximum problem. Cohler was an early employee at Facebook, and he's widely considered one of Silicon Valley's smartest thinkers on the social Web.

The local-maximum problem, he explains to me, shows up any time you're trying to optimize something. Say you're trying to write a simple set of instructions to help a blind person who's lost in the Sierra Nevadas find his way to the highest peak. "Feel around you to see if you're surrounded by downward-sloping land," you say. "If you're not, move in a direction that's higher, and repeat."

Programmers face problems like this all the time. What link is the best result for the search term "fish"? Which picture can Facebook show you to increase the likelihood that you'll start a photo-surfing binge? The directions sound pretty obvious— you just tweak and tune in one direction or another until you're in the sweet spot. But there's a problem with these hill-climbing instructions: They're as likely to end you up in the foothills— the local maximum—as they are to guide you to the apex of Mount Whitney.

This isn't exactly harmful, but in the filter bubble, the same phenomenon can happen with any person or topic. I find it hard not to click on articles about gadgets, though I don't actually think they're that important. Personalized filters play to the most compulsive parts of you, creating "compulsive media" to get you to click things more. The technology mostly can't distinguish compulsion from general interest—and if you're generating page views that can be sold to advertisers, it might not care.

The faster the system learns from you, the more likely it is that you can get trapped in a kind of identity cascade, in which a small initial action—clicking on a link about gardening or anarchy or Ozzy Osbourne—indicates that you're a person who likes those kinds of things. This in turn supplies you with more information on the topic, which you're more inclined to click on because the topic has now been primed for you.

Especially once the second click has occurred, your brain is in on the act as well. Our brains act to reduce cognitive dissonance in a strange but compelling kind of unlogic—"Why would I have done x if I weren't a person who does x—therefore I must be a person who does x." Each click you take in this loop is another action to self-justify—"Boy, I guess I just really love 'Crazy Train.'" When you use a recursive process that feeds on itself, Cohler tells me, "You're going to end up down a deep and narrow path." The reverb drowns out the tune. If identity loops aren't counteracted through randomness and serendipity, you could end up stuck in the foothills of your identity, far away from the high peaks in the distance.

And that's when these loops are relatively benign. Sometimes they're not.

We know what happens when teachers think students are dumb: They get dumber. In an experiment done before the advent of ethics boards, teachers were given test results that supposedly indicated the IQ and aptitude of students entering their classes. They weren't told, however, that the results had been randomly redistributed among students. After a year, the students who the teachers had been told were bright made big gains in IQ. The students who the teachers had been told were below average had no such improvement.

So what happens when the Internet thinks you're dumb? Personalization based on perceived IQ isn't such a far-fetched scenario—Google Docs even offers a helpful tool for automatically checking the grade-level of written text. If your education level isn't already available through a tool like Acxiom, it's easy enough for anyone with access to a few e-mails or Facebook posts to infer. Users whose writing indicates college-level literacy might see more articles from the *New Yorker;* users with only basic writing skills might see more from the *New York Post.*

In a broadcast world, everyone is expected to read or process information at about the same level. In the filter bubble, there's no need for that expectation. On one hand, this could be great—vast groups of people who have given up on reading because the newspaper goes over their heads may finally connect with written content. But without pressure to improve, it's also possible to get stuck in a grade-three world for a long time.

Incidents and Adventures

In some cases, letting algorithms make decisions about what we see and what opportunities we're offered gives us fairer results.

A computer can be made blind to race and gender in ways that humans usually can't. But that's only if the relevant algorithms are designed with care and acuteness. Otherwise, they're likely to simply reflect the social mores of the culture they're processing—a regression to the social norm.

In some cases, algorithmic sorting based on personal data can be even *more* discriminatory than people would be. For example, software that helps companies sift through résumés for talent might "learn" by looking at which of its recommended employees are actually hired. If nine white candidates in a row are chosen, it might determine that the company isn't interested in hiring black people and exclude them from future searches. "In many ways," writes NYU sociologist Dalton Conley, "such network-based categorizations are more insidious than the hackneyed groupings based on race, class, gender, religion, or any other demographic characteristic." Among programmers, this kind of error has a name. It's called *overfitting*.

The online movie rental Web site Netflix is powered by an algorithm called CineMatch. To start, it was pretty simple. If I had rented the first movie in the *Lord of the Rings* trilogy, let's say, Netflix could look up what other movies *Lord of the Rings* watchers had rented. If many of them had rented *Star Wars*, it'd be highly likely that I would want to rent it, too.

This technique is called kNN (k-nearest-neighbor), and using it CineMatch got pretty good at figuring out what movies people wanted to watch based on what movies they'd rented and how many stars (out of five) they'd given the movies they'd seen. By 2006, CineMatch could predict within one star how much a given user would like any movie from Netflix's vast

hundred-thousand-film emporium. Already CineMatch was better at making recommendations than most humans. A human
video clerk would never think to suggest *Silence of the Lambs*
to a fan of *The Wizard of Oz*, but CineMatch knew people
who liked one usually liked the other.

But Reed Hastings, Netflix's CEO, wasn't satisfied. "Right
now, we're driving the Model-T version of what's possible," he
told a reporter in 2006. On October 2, 2006, an announcement
went up on the Netflix Web site: "We're interested, to the tune
of $1 million." Netflix had posted an enormous swath of data—
reviews, rental records, and other information from its user
database, scrubbed of anything that would obviously identify a
specific user. And now the company was willing to give $1 mil-
lion to the person or team who beat CineMatch by more than
10 percent. Like the longitude prize, the Netflix Challenge was
open to everyone. "All you need is a PC and some great insight,"
Hastings declared in the *New York Times*.

After nine months, about eighteen thousand teams from
more than 150 countries were competing, using ideas from
machine learning, neural networks, collaborative filtering, and
data mining. Usually, contestants in high-stakes contests operate in secret. But Netflix encouraged the competing groups to
communicate with one another and built a message board
where they could coordinate around common obstacles. Read
through the message board, and you get a visceral sense of the
challenges that bedeviled the contestants during the three-year
quest for a better algorithm. Overfitting comes up again and
again.

There are two challenges in building pattern-finding algorithms. One is finding the patterns that are there in all the noise. The other problem is the opposite: *not* finding patterns in the data that aren't actually really there. The pattern that describes "1, 2, 3" could be "add one to the previous number" or "list positive prime numbers from smallest to biggest." You don't know for sure until you get more data. And if you leap to conclusions, you're overfitting.

Where movies are concerned, the dangers of overfitting are relatively small—many analog movie watchers have been led to believe that because they liked *The Godfather* and *The Godfather: Part II*, they'll like *The Godfather: Part III*. But the overfitting problem gets to one of the central, irreducible problems of the filter bubble: Overfitting and stereotyping are synonyms.

The term *stereotyping* (which in this sense comes from Walter Lippmann, incidentally) is often used to refer to malicious xenophobic patterns that aren't true—"people of this skin color are less intelligent" is a classic example. But stereotypes and the negative consequences that flow from them aren't fair to specific people *even if* they're generally pretty accurate.

Marketers are already exploring the gray area between what can be predicted and what predictions are fair. According to Charlie Stryker, an old hand in the behavioral targeting industry who spoke at the Social Graph Symposium, the U.S. Army has had terrific success using social-graph data to recruit for the military—after all, if six of your Facebook buddies have enlisted, it's likely that you would consider doing so too. Drawing inferences based on what people like you or people linked to you do is pretty good business. And it's not just the army. Banks are beginning to use social data to decide to whom to offer loans:

If your friends don't pay on time, it's likely that you'll be a deadbeat too. "A decision is going to be made on creditworthiness based on the creditworthiness of your friends," Stryker said. "There are applications of this technology that can be very powerful," another social targeting entrepreneur told the *Wall Street Journal*. "Who knows how far we'd take it?"

Part of what's troubling about this world is that companies aren't required to explain on what basis they're making these decisions. And as a result, you can get judged without knowing it and without being able to appeal. For example, LinkedIn, the social job-hunting site, offers a career trajectory prediction site; by comparing your résumé to other peoples' who are in your field but further along, LinkedIn can forecast where you'll be in five years. Engineers at the company hope that soon it'll be able to pinpoint career choices that lead to better outcomes—"mid-level IT professionals like you who attended Wharton business school made $25,000/year more than those who didn't." As a service to customers, it's pretty useful. But imagine if LinkedIn provided that data to corporate clients to help them weed out people who are forecast to be losers. Because that could happen entirely without your knowledge, you'd never get the chance to argue, to prove the prediction wrong, to have the benefit of the doubt.

If it seems unfair for banks to discriminate against you because your high school buddy is bad at paying his bills or because you like something that a lot of loan defaulters also like, well, it is. And it points to a basic problem with induction, the logical method by which algorithms use data to make predictions.

Philosophers have been wrestling with this problem since long before there were computers to induce with. While you can prove the truth of a mathematical proof by arguing it out from first principles, the philosopher David Hume pointed out in 1772 that reality doesn't work that way. As the investment cliché has it, past performance is not indicative of future results.

This raises some big questions for science, which is at its core a method for using data to predict the future. Karl Popper, one of the preeminent philosophers of science, made it his life's mission to try to sort out the problem of induction, as it came to be known. While the optimistic thinkers of the late 1800s looked at the history of science and saw a journey toward truth, Popper preferred to focus on the wreckage along the side of the road—the abundance of failed theories and ideas that were perfectly consistent with the scientific method and yet horribly wrong. After all, the Ptolemaic universe, with the earth in the center and the sun and planets revolving around it, survived an awful lot of mathematical scrutiny and scientific observation.

Popper posed his problem in a slightly different way: Just because you've only ever seen white swans doesn't mean that all swans are white. What you have to look for is the black swan, the counterexample that proves the theory wrong. "Falsifiability," Popper argued, was the key to the search for truth: The purpose of science, for Popper, was to advance the biggest claims for which one could not find any countervailing examples, any black swans. Underlying Popper's view was a deep humility about scientifically induced knowledge—a sense that we're wrong as often as we're right, and we usually don't know when we are.

It's this humility that many algorithmic prediction methods fail to build in. Sure, they encounter people or behaviors that don't fit the mold from time to time, but these aberrations don't fundamentally compromise their algorithms. After all, the advertisers whose money drives these systems don't need the models to be perfect. They're most interested in hitting demographics, not complex human beings.

When you model the weather and predict there's a 70 percent chance of rain, it doesn't affect the rain clouds. It either rains or it doesn't. But when you predict that because my friends are untrustworthy, there's a 70 percent chance that I'll default on my loan, there are consequences if you get me wrong. You're discriminating.

The best way to avoid overfitting, as Popper suggests, is to try to prove the model wrong and to build algorithms that give the benefit of the doubt. If Netflix shows me a romantic comedy and I like it, it'll show me another one and begin to think of me as a romantic-comedy lover. But if it wants to get a good picture of who I really am, it should be constantly testing the hypothesis by showing me *Blade Runner* in an attempt to prove it wrong. Otherwise, I end up caught in a local maximum populated by Hugh Grant and Julia Roberts.

The statistical models that make up the filter bubble write off the outliers. But in human life it's the outliers who make things interesting and give us inspiration. And it's the outliers who are the first signs of change.

One of the best critiques of algorithmic prediction comes, remarkably, from the late-nineteenth-century Russian novelist Fyodor Dostoyevsky, whose *Notes from Underground* was a passionate critique of the utopian scientific rationalism of the day. Dostoyevsky looked at the regimented, ordered human life that science promised and predicted a banal future. "All human actions," the novel's unnamed narrator grumbles, "will then, of course, be tabulated according to these laws, mathematically, like tables of logarithms up to 108,000, and entered in an index . . . in which everything will be so clearly calculated and explained that there will be no more incidents or adventures in the world."

The world often follows predictable rules and falls into predictable patterns: Tides rise and fall, eclipses approach and pass; even the weather is more and more predictable. But when this way of thinking is applied to human behavior, it can be dangerous, for the simple reason that our best moments are often the most unpredictable ones. An entirely predictable life isn't worth living. But algorithmic induction can lead to a kind of information determinism, in which our past clickstreams entirely decide our future. If we don't erase our Web histories, in other words, we may be doomed to repeat them.

The Public Is Irrelevant

The presence of others who see what we see and
hear what we hear assures us of the reality of the
world and ourselves.

—Hannah Arendt

It is an axiom of political science in the United
States that the only way to neutralize the influence
of the newspapers is to multiply their number.

—Alexis de Tocqueville

On the night of May 7, 1999, a B-2 stealth bomber left
Whiteman Air Force Base in Missouri. The aircraft
flew on an easterly course until it reached the city of
Belgrade in Serbia, where a civil war was under way. Around
midnight local time, the bomber delivered its cargo: four GPS-
guided bombs, into which had been programmed an address
that CIA documents identified as a possible arms warehouse. In
fact, the address was the Yugoslavian Chinese Embassy. The
building was demolished, and three Chinese diplomats were
killed.

The United States immediately apologized, calling the event

an accident. On Chinese state TV, however, an official statement called the bombing a "barbaric attack and a gross violation of Chinese sovereignty." Though President Bill Clinton tried to reach Chinese President Jiang Zemin, Zemin repeatedly rejected his calls; Clinton's videotaped apology to the Chinese people was barred from Chinese media for four days.

As anti-U.S. riots began to break out in the streets, China's largest newspaper, the *People's Daily*, created an online chat forum called the Anti-Bombing Forum. Already, in 1999, chat forums were huge in China—much larger than they've ever been in the United States. As *New York Times* journalist Tom Downey explained a few years later, "News sites and individual blogs aren't nearly as influential in China, and social networking hasn't really taken off. What remain most vital are the largely anonymous online forums . . . that are much more participatory, dynamic, populist and perhaps even democratic than anything on the English-language Internet." Tech writer Clive Thompson quotes Shanthi Kalathil, a researcher at the Carnegie Endowment, who says that the Anti-Bombing Forum helped to legitimize the Chinese government's position that the bombing was deliberate among "an elite, wired section of the population." The forum was a form of crowd-sourced propaganda: Rather than just telling Chinese citizens what to think, it lifted the voices of thousands of patriots aligned with the state.

Most of the Western reporting on Chinese information management focuses on censorship: Google's choice to remove, temporarily, search results for "Tiananmen Square," or Microsoft's decision to ban the word "democracy" from Chinese blog posts, or the Great Firewall that sits between China and the outside world and sifts through every packet of information that enters or exits the country. Censorship in China is real: There are plenty of words that have been more or less stricken from the public discourse. When Thompson asks whether the popular Alibaba engine would show results for dissident movements, CEO Jack Ma shook his head. "No! We are a business!" he said. "Shareholders want to make money. Shareholders want us to make the customer happy. Meanwhile we do not have any responsibilities saying we should do this or that political thing."

In practice, the firewall is not so hard to circumvent. Corporate virtual private networks—Internet connections encrypted to prevent espionage—operate with impunity. Proxies and firewall workarounds like Tor connect in-country Chinese dissidents with even the most hard-core antigovernment Web sites. But to focus exclusively on the firewall's inability to perfectly block information is to miss the point. China's objective isn't so much to blot out unsavory information as to alter the physics around it—to create friction for problematic information and to route public attention to progovernment forums. While it can't block all of the people from all of the news all of the time, it doesn't need to.

"What the government cares about," *Atlantic* journalist James Fallows writes, "is making the quest for information just enough of a nuisance that people generally won't bother." The strategy, says Xiao Qiang of the University of California at Berkeley, is "about social control, human surveillance, peer pressure, and self-censorship." Because there's no official list of blocked keywords or forbidden topics published by the government, businesses and individuals censor themselves to avoid a visit from the police. Which sites are available changes daily. And while some bloggers suggest that the system's unreliability is a result of faulty technology ("the Internet will override attempts to control it!"), for the government this is a feature, not a bug. James Mulvenon, the head of the Center for Intelligence Research and Analysis, puts it this way: "There's a randomness to their enforcement, and that creates a sense that they're looking at everything."

Lest that sensation be too subtle, the Public Security Bureau in Shenzhen, China, developed a more direct approach: Jingjing and Chacha, the cartoon Internet Police. As the director of the initiative told the *China Digital Times*, he wanted "to let all Internet users know that the Internet is not a place beyond law [and that] the Internet Police will maintain order in all online behavior." Icons of the male-female pair, complete with jaunty flying epaulets and smart black shoes, were placed on all major Web sites in Shenzhen; they even had instant-message addresses so that six police officers could field questions from the online crowds.

"People are actually quite free to talk about [democracy]," Google's China point man, Kai-Fu Lee, told Thompson in 2006. "I don't think they care that much. Hey, U.S. democracy, that's a good form of government. Chinese government, good and stable, that's a good form of government. Whatever, as long as I get to go to my favorite Web site, see my friends, live happily." It may not be a coincidence that the Great Firewall stopped blocking pornography recently. "Maybe they are thinking that if Internet users have some porn to look at, then they won't pay so much attention to political matters," Michael Anti, a Beijing-based analyst, told the AP.

We usually think about censorship as a process by which governments alter facts and content. When the Internet came along, many hoped it would eliminate censorship altogether—the flow of information would simply be too swift and strong for governments to control. "There's no question China has been trying to crack down on the Internet," Bill Clinton told the audience at a March 2000 speech at Johns Hopkins University. "Good luck! That's sort of like trying to nail Jell-O to the wall."

But in the age of the Internet, it's still possible for governments to manipulate the truth. The process has just taken a different shape: Rather than simply banning certain words or opinions outright, it'll increasingly revolve around second-order censorship—the manipulation of curation, context, and the flow of information and attention. And because the filter bubble is primarily controlled by a few centralized companies, it's not as difficult to adjust this flow on an individual-by-individual basis as you might think. Rather than decentralizing power, as its early proponents predicted, in some ways the Internet is concentrating it.

Lords of the Cloud

To get a sense of how personalization might be used for political ends, I talked to a man named John Rendon.

Rendon affably describes himself as an "information warrior and perception manager." From the Rendon Group's headquarters in Washington, D.C.'s, Dupont Circle, he provides those services to dozens of U.S. agencies and foreign governments. When American troops rolled into Kuwait City during the first Iraq war, television cameras captured hundreds of Kuwaitis joyfully waving American flags. "Did you ever stop to wonder," he asked an audience later, "how the people of Kuwait City, after being held hostage for seven long and painful months, were able to get handheld American flags? And for that matter, the flags of other coalition countries? Well, you now know the answer. That was one of my jobs."

Much of Rendon's work is confidential—he enjoys a level of beyond–Top Secret clearance that even high-level intelligence analysts sometimes fail to get. His role in George W. Bush–era pro-U.S. propaganda in Iraq is unclear: While some sources claim he was a central figure in the effort, Rendon denies any involvement. But his dream is quite clear: Rendon wants to see a world where television "can drive the policy process," where "border patrols [are] replaced by beaming patrols," and where "you can win without fighting."

Given all that, I was a bit surprised when the first weapon he referred me to was a very quotidian one: a thesaurus. The key to changing public opinion, Rendon said, is finding different ways to say the same thing. He described a matrix, with extreme language or opinion on one side and mild opinion on the other. By using sentiment analysis to figure out how people in a country felt about an event—say, a new arms deal with the United States—and identify the right synonyms to move them toward approval, you could "gradually nudge a debate." "It's a lot easier to be close to what reality is" and push it in the right direction, he said, than to make up a new reality entirely.

Rendon had seen me talk about personalization at an event we both attended. Filter bubbles, he told me, provided new ways of managing perceptions. "It begins with getting inside the algorithm. If you could find a way to load your content up so that only your content gets pulled by the stalking algorithm, then you'd have a better chance of shaping belief sets," he said.

In fact, he suggested, if we looked in the right places, we might be able to see traces of this kind of thing happening now—sentiment being algorithmically shifted over time.

But if the filter bubble might make shifting perspectives easier in a future Iraq or Panama, Rendon was clearly concerned about the impact of self-sorting and personalized filtering for democracy at home. "If I'm taking a photo of a tree," he said, "I need to know what season we're in. Every season it looks different. It could be dying, or just losing its leaves in autumn." To make good decisions, context is crucial—that's why the military is so focused on what they call "360-degree situational awareness." In the filter bubble, you don't get 360 degrees—and you might not get more than one.

I returned to the question about using algorithms to shift sentiment. "How does someone game the system when it's all about self-generated, self-reinforcing information flows? I have to think about it more," Rendon said, "But I think I know how I'd do it."

"How?" I asked.

He paused, then chuckled: "Nice try." He'd already said too much.

The campaign of propaganda that Walter Lippmann railed against in World War I was a massive undertaking: To "goose-step the truth," hundreds of newspapers nationwide had to be brought onboard. Now that every blogger is a publisher, the task seems nearly impossible. In 2010, Google chief Eric Schmidt echoed this sentiment, arguing in the journal *Foreign Affairs* that the Internet eclipses intermediaries and governments and empowers individuals to "consume, distribute, and create their own content without government control."

It's a convenient view for Google—if intermediaries are losing power, then the company's merely a minor player in a much larger drama. But in practice, a great majority of online content reaches people through a small number of Web sites—Google foremost among them. These big companies represent new loci of power. And while their multinational character makes them resistant to some forms of regulation, they can also offer one-stop shopping for governments seeking to influence information flows.

As long as a database exists, it's potentially accessible by the state. That's why gun rights activists talk a lot about Alfred Flatow. Flatow was an Olympic gymnast and German Jew who in 1932 registered his gun in accordance with the laws of the waning Weimar Republic. In 1938, German police came to his door. They'd searched through the record, and in preparation for the Holocaust, they were rounding up Jews with handguns. Flatow was killed in a concentration camp in 1942.

For National Rifle Association members, the story is a powerful cautionary tale about the dangers of a national gun registry. As a result of Flatow's story and thousands like it, the NRA has successfully blocked a national gun registry for decades. If a fascistic anti-Semitic regime came into power in the United States, it'd be hard put to identify gun-holding Jews using its own databases.

But the NRA's focus may have been too narrow. Fascists aren't known for carefully following the letter of the law regarding extragovernmental databases. And using the data that credit card companies use—or for that matter, building models based on the thousands of data points Acxiom tracks—it'd be a simple matter to predict with significant accuracy who has a gun and who does not.

Even if you aren't a gun advocate, the story is worth paying attention to. The dynamics of personalization shift power into the hands of a few major corporate actors. And this consolidation of huge masses of data offers governments (even democratic ones) more potential power than ever.

Rather than housing their Web sites and databases internally, many businesses and start-ups now run on virtual computers in vast server farms managed by other companies. The enormous pool of computing power and storage these networked machines create is known as the *cloud*, and it allows clients much greater flexibility. If your business runs in the cloud, you don't need to buy more hardware when your processing demands expand: You just rent a greater portion of the cloud. Amazon Web Services, one of the major players in the space, hosts thousands of Web sites and Web servers and undoubtedly stores the personal data of millions. On one hand, the cloud gives every kid in his or her basement access to nearly unlimited computing power to quickly scale up a new online service. On the other, as Clive Thompson pointed out to me, the cloud "is actually just a handful of companies." When Amazon booted the activist Web site WikiLeaks off its servers under political pressure in 2010, the site immediately collapsed—there was nowhere to go.

Personal data stored in the cloud is also actually much easier for the government to search than information on a home computer. The FBI needs a warrant from a judge to search your laptop. But if you use Yahoo or Gmail or Hotmail for your e-mail, you "lose your constitutional protections immediately," according to a lawyer for the Electronic Freedom Foundation. The FBI can just ask the company for the information—no judicial paperwork needed, no permission required—as long as it can argue later that it's part of an "emergency." "The cops will love this," says privacy advocate Robert Gellman about cloud computing. "They can go to a single place and get everybody's documents."

Because of the economies of scale in data, the cloud giants are increasingly powerful. And because they're so susceptible to regulation, these companies have a vested interest in keeping government entities happy. When the Justice Department requested billions of search records from AOL, Yahoo, and MSN in 2006, the three companies quickly complied. (Google, to its credit, opted to fight the request.) Stephen Arnold, an IT expert who worked at consulting firm Booz Allen Hamilton, says that Google at one point housed three officers of "an unnamed intelligence agency" at its headquarters in Mountain View. And Google and the CIA have invested together in a firm called Recorded Future, which focuses on using data connections to predict future real-world events.

Even if the consolidation of this data-power doesn't result in more governmental control, it's worrisome on its own terms.

One of the defining traits of the new personal information environment is that it's asymmetrical. As Jonathan Zittrain argues in *The Future of the Internet—and How to Stop It*, "nowadays, an individual must increasingly give information about himself to large and relatively faceless institutions, for handling and use by strangers—unknown, unseen, and all too frequently, unresponsive."

In a small town or an apartment building with paper-thin walls, what I know about you is roughly the same as what you know about me. That's a basis for a social contract, in which we'll deliberately ignore some of what we know. The new privacyless world does away with that contract. I can know a lot about you without your knowing I know. "There's an implicit bargain in our behavior," search expert John Battelle told me, "that we haven't done the math on."

If Sir Francis Bacon is right that "knowledge is power," privacy proponent Viktor Mayer-Schonberger writes that what we're witnessing now is nothing less than a "redistribution of information power from the powerless to the powerful." It'd be one thing if we all knew everything about each other. It's another when centralized entities know a lot more about us than we know about each other—and sometimes, more than we know about ourselves. If knowledge is power, then asymmetries in knowledge are asymmetries in power.

Google's famous "Don't be evil" motto is presumably intended to allay some of these concerns. I once explained to a Google search engineer that while I didn't think the company was currently evil, it seemed to have at its fingertips everything it needed to do evil if it wished. He smiled broadly. "Right," he said. "We're not evil. We try really hard not to be evil. But if we wanted to, man, could we ever!"

Friendly World Syndrome

Most governments and corporations have used the new power that personal data and personalization offer fairly cautiously so far—China, Iran, and other oppressive regimes being the obvious exceptions. But even putting aside intentional manipulation, the rise of filtering has a number of unintended yet serious consequences for democracies. In the filter bubble, the public sphere—the realm in which common problems are identified and addressed—is just less relevant.

For one thing, there's the problem of the friendly world.

Communications researcher George Gerbner was one of the first theorists to look into how media affect our political beliefs, and in the mid-1970s, he spent a lot of time thinking about shows like *Starsky and Hutch*. It was a pretty silly program, filled with the shared clichés of seventies cop TV—the bushy moustaches, the twanging soundtracks, the simplistic good-versus-evil plots. And it was hardly the only one—for every *Charlie's Angels* or *Hawaii Five-O* that earned a place in cultural memory, there are dozens of shows, like *The Rockford Files*, *Get Christie Love*, and *Adam-12*, that are unlikely to be resuscitated for ironic twenty-first-century remakes.

But Gerbner, a World War II veteran–turned–communications theorist who became dean of the Annenberg School of Communication, took these shows seriously. Starting in 1969, he began a systematic study of the way TV programming affects how we think about the world. As it turned out, the *Starsky and Hutch* effect was significant. When you asked TV watchers to estimate the percentage of the adult workforce that was made up of cops, they vastly overguessed relative to non–TV watchers with the same education and demographic background. Even more troubling, kids who saw a lot of TV violence were much more likely to be worried about real-world violence.

Gerbner called this the mean world syndrome: If you grow up in a home where there's more than, say, three hours of television per day, for all practical purposes, you live in a meaner world—and act accordingly—than your next-door neighbor who lives in the same place but watches less television. "You know, who tells the stories of a culture really governs human behavior," Gerbner later said.

Gerbner died in 2005, but he lived long enough to see the Internet begin to break that stranglehold. It must have been a relief: Although our online cultural storytellers are still quite consolidated, the Internet at least offers more choice. If you want to get your local news from a blogger rather than a local TV station that trumpets crime rates to get ratings, you can.

But if the mean world syndrome poses less of a risk these days, there's a new problem on the horizon: We may now face what persuasion-profiling theorist Dean Eckles calls a friendly world syndrome, in which some of the biggest and most important problems fail to reach our view at all.

While the mean world on television arises from a cynical "if it bleeds, it leads" approach to programming, the friendly world generated by algorithmic filtering may not be as intentional. According to Facebook engineer Andrew Bosworth, the team that developed the Like button originally considered a number of options—from stars to a thumbs up sign (but in Iran and Thailand, it's an obscene gesture). For a month in the summer of 2007, the button was known as the Awesome button. Eventually, however, the Facebook team gravitated toward Like, which is more universal.

That Facebook chose Like instead of, say, Important is a small design decision with far-reaching consequences: The stories that get the most attention on Facebook are the stories that get the most Likes, and the stories that get the most Likes are, well, more likable.

Facebook is hardly the only filtering service that will tend toward an antiseptically friendly world. As Eckles pointed out to me, even Twitter, which has a reputation for putting filtering in the hands of its users, has this tendency. Twitter users see most of the tweets of the folks they follow, but if my friend is having an exchange with someone I don't follow, it doesn't show up. The intent is entirely innocuous: Twitter is trying not to inundate me with conversations I'm not interested in. But the result is that conversations between my friends (who will tend to be like me) are overrepresented, while conversations that could introduce me to new ideas are obscured.

Of course, *friendly* doesn't describe all of the stories that pierce the filter bubble and shape our sense of the political world. As a progressive political news junkie, I get plenty of news about Sarah Palin and Glenn Beck. The valence of this news, however, is very predictable: People are posting it to signal their dismay with Beck's and Palin's rhetoric and to build a sense of solidarity with their friends, who presumably feel the same way. It's rare that my assumptions about the world are shaken by what I see in my news feed.

Emotional stories are the ones that generally thrive in the filter bubble. The Wharton School study on the *New York Times*'s Most Forwarded List, discussed in chapter 2, found that stories that aroused strong feelings—awe, anxiety, anger, happiness—were much more likely to be shared. If television gives us a "mean world," filter bubbles give us an "emotional world."

One of the troubling side effects of the friendly world

syndrome is that some important public problems will disappear. Few people seek out information about homelessness, or share it, for that matter. In general, dry, complex, slow-moving problems—a lot of the truly significant issues—won't make the cut. And while we used to rely on human editors to spotlight these crucial problems, their influence is now waning.

Even advertising isn't necessarily a foolproof way of alerting people to public problems, as the environmental group Oceana found out. In 2004, Oceana was running a campaign urging Royal Caribbean to stop dumping its raw sewage into the sea; as part of the campaign, it took out a Google ad that said "Help us protect the world's oceans. Join the fight!" After two days, Google pulled the ads, citing "language advocating against the cruise line industry" that was in violation of their general guidelines about taste. Apparently, advertisers that implicated corporations in public issues weren't welcome.

The filter bubble will often block out the things in our society that are important but complex or unpleasant. It renders them invisible. And it's not just the issues that disappear. Increasingly, it's the whole political process.

The Invisible Campaign

When George W. Bush came out of the 2000 election with far fewer votes than Karl Rove expected, Rove set in motion a series of experiments in microtargeted media in Georgia—looking at a wide range of consumer data ("Do you prefer beer or wine?") to try to predict voting behavior and identify who was persuadable and who could be easily motivated to get to the polls. Though the findings are still secret, legend has it that the methods Rove discovered were at the heart of the GOP's successful get-out-the-vote strategy in 2002 and 2004.

On the left, Catalist, a firm staffed by former Amazon engineers, has built a database of hundreds of millions of voter profiles. For a fee, organizing and activist groups (including MoveOn) query it to help determine which doors to knock on and to whom to run ads. And that's just the start. In a memo for fellow progressives, Mark Steitz, one of the primary Democratic data gurus, recently wrote that "targeting too often returns to a bombing metaphor—dropping message from planes. Yet the best data tools help build relationships based on observed contacts with people. Someone at the door finds out someone is interested in education; we get back to that person and others like him or her with more information. Amazon's recommendation engine is the direction we need to head." The trend is clear: We're moving from swing states to swing people.

Consider this scenario: It's 2016, and the race is on for the presidency of the United States. Or is it?

It depends on who you are, really. If the data says you vote frequently and that you may have been a swing voter in the past, the race is a maelstrom. You're besieged with ads, calls, and invitations from friends. If you vote intermittently, you get a lot of encouragement to get out to the polls.

But let's say you're more like an average American. You usually vote for candidates from one party. To the data crunchers from the opposing party, you don't look particularly persuadable. And because you vote in presidential elections pretty regularly, you're also not a target for "get out the vote" calls from your own. Though you make it to the polls as a matter of civic duty, you're not that actively interested in politics. You're more interested in, say, soccer and robots and curing cancer and what's going on in the town where you live. Your personalized news feeds reflect those interests, not the news from the latest campaign stop.

In a filtered world, with candidates microtargeting the few persuadables, would you know that the campaign was happening at all?

Even if you visit a site that aims to cover the race for a general audience, it'll be difficult to tell what's going on. What is the campaign about? There is no general, top-line message, because the candidates aren't appealing to a general public. Instead, there are a series of message fragments designed to penetrate personalized filters.

Google is preparing for this future. Even in 2010, it staffed a round-the-clock "war room" for political advertising, aiming to be able to quickly sign off on and activate new ads even in the wee hours of October nights. Yahoo is conducting a series of experiments to determine how to match the publicly available list of who voted in each district with the click signals and Web history data it picks up on its site. And data-aggregation firms like Rapleaf in San Francisco are trying to correlate Facebook social graph information with voting behavior—so that they can show you the political ad that best works for you based on the responses of your friends.

The impulse to talk to voters about the things they're actually interested in isn't a bad one—it'd be great if mere mention of the word *politics* didn't cause so many eyes to glaze over. And certainly the Internet has unleashed the coordinated energy of a whole new generation of activists—it's easier than ever to find people who share your political passions. But while it's easier than ever to bring a group of people together, as personalization advances it'll become harder for any given group to reach a broad audience. In some ways, personalization poses a threat to public life itself.

Because the state of the art in political advertising is half a decade behind the state of the art in commercial advertising, most of this change is still to come. But for starters, filter-bubble politics could effectively make even more of us into single-issue voters. Like personalized media, personalized advertising is a two-way street: I may see an ad about, say, preserving the environment because I drive a Prius, but seeing the ad also makes me care more about preserving the environment. And if a congressional campaign can determine that this is the issue on which it's most likely to persuade me, why bother filling me in on all of the other issues?

In theory, market dynamics will continue to encourage campaigns to reach out to nonvoters. But an additional complication is that more and more companies are also allowing users to remove advertisements they don't like. For Facebook and Google, after all, seeing ads for ideas or services you don't like is a failure. Because people tend to dislike ads containing messages they disagree with, this creates even less space for persuasion. "If a certain number of anti-Mitt Republicans saw an ad for Mitt Romney and clicked 'offensive, etc.,'" writes Vincent Harris, a Republican political consultant, "they could block ALL of Mitt Romney's ads from being shown, and kill the entire online advertising campaign regardless of how much money the Romney campaign wanted to spend on Facebook." Forcing candidates to come up with more palatable ways to make their points might result in more thoughtful ads—but it also might also drive up the cost of these ads, making it too costly for campaigns to ever engage the other side.

The most serious political problem posed by filter bubbles is that they make it increasingly difficult to have a public argument. As the number of different segments and messages increases, it becomes harder and harder for the campaigns to track who's saying what to whom. TV is a piece of cake to monitor in comparison—you can just record the opposition's ads in each cable district. But how does a campaign know what its opponent is saying if ads are only targeted to white Jewish men between twenty-eight and thirty-four who have expressed a fondness for U2 on Facebook and who donated to Barack Obama's campaign?

When a conservative political group called Americans for Job Security ran ads in 2010 falsely accusing Representative Pete Hoekstra of refusing to sign a no-new-taxes pledge, he was able to show TV stations the signed pledge and have the ads pulled off the air. It's not great to have TV station owners be the sole arbitrators of truth—I've spent a fair amount of time arguing with them myself—but it is better to have some bar for truthfulness than none at all. It's unclear that companies like Google have the resources or the interest to play truthfulness referee on the hundreds of thousands of different ads that will run in election cycles to come.

As personal political targeting increases, not only will it be

more difficult for campaigns to respond to and fact-check each
other, it'll be more challenging for journalists as well. We may
see an environment where the most important ads aren't easily
accessible to journalists and bloggers—it's easy enough for
campaigns to exclude them from their targeting and difficult
for reporters to fabricate the profile of a genuine swing voter.
(One simple solution to this problem would simply be to
require campaigns to immediately disclose all of their online
advertising materials and to whom each ad is targeted. Right
now, the former is spotty and the latter is undisclosed.)

It's not that political TV ads are so great. For the most part,
they're shrill, unpleasant, and unlikable. If we could, most of us
would tune them out. But in the broadcast era, they did at least
three useful things. They reminded people that there was an
election in the first place. They established for everyone what
the candidates valued, what their campaigns were about, what
their arguments were: the parameters of the debate. And they
provided a basis for a common conversation about the political
decision we faced—something you could talk about in the line
at the supermarket.

For all of their faults, political campaigns are one of the primary places where we debate our ideas about our nation. Does
America condone torture? Are we a nation of social Darwinists
or of social welfare? Who are our heroes, and who are our vil-
lains? In the broadcast era, campaigns have helped to delineate
the answers to those questions. But they may not do so for very
much longer.

Fragmentation

The aim of modern political marketing, consumer trends expert J. Walker Smith tells Bill Bishop in *The Big Sort*, is to "drive customer loyalty—and in marketing terms, drive the average transaction size or improve the likelihood that a registered Republican will get out and vote Republican. That's a business philosophy applied to politics that I think is really dangerous, because it's not about trying to form a consensus, to get people to think about the greater good."

In part, this approach to politics is on the rise for the same reason the filter bubble is: Personalized outreach gives better bang for the political buck. But it's also a natural outcome of a well-documented shift in how people in industrialized countries think about what's important. When people don't have to worry about having their basic needs met, they care a lot more about having products and leaders that represent who they are.

Professor Ron Inglehart calls this trend postmaterialism, and it's a result of the basic premise, he writes, that "you place the greatest subjective value on the things in short supply." In surveys spanning forty years and eighty countries, people who were raised in affluence—who never had to worry about their physical survival—behaved in ways strikingly different from those of their hungry parents. "We can even specify," Inglehart writes in *Modernization and Postmodernization*, "with far better than random success, what issues are likely to be most salient in the politics of the respective types of societies."

While there are still significant differences from country to country, postmaterialists share some important traits. They're less reverent about authority and traditional institutions—the appeal of authoritarian strongmen appears to be connected to a basic fear for survival. They're more tolerant of difference: One especially striking chart shows a strong correlation between level of life satisfaction and comfort with living next door to someone who's gay. And while earlier generations emphasize financial achievement and order, postmaterialists value self-expression and "being yourself."

Somewhat confusingly, postmaterialism doesn't mean anti-consumption. Actually, the phenomenon is at the bedrock of our current consumer culture: Whereas we once bought things because we needed them to survive, now we mostly buy things as a means of self-expression. And the same dynamics hold for political leadership: Increasingly, voters evaluate candidates on whether they represent an aspirational version of themselves.

The result is what marketers call brand fragmentation. When brands were primarily about validating the quality of a product—"Dove soap is pure and made of the best ingredients"—advertisements focused more on the basic value proposition. But when brands became vehicles for expressing identity, they needed to speak more intimately to different groups of people with divergent identities they wanted express. And as a result, they started to splinter. Which is why what's happened to Pabst Blue Ribbon beer is a good way of understanding the challenges faced by Barack Obama.

In the early 2000s, Pabst was struggling financially. It had maxed out among the white rural population that formed the core of its customer base, and it was selling less than 1 million barrels of beer a year, down from 20 million in 1970. If Pabst wanted to sell more beer, it had to look elsewhere, and Neal Stewart, a midlevel marketing manager, did. Stewart went to Portland, Oregon, where Pabst numbers were surprisingly strong and an ironic nostalgia for white working-class culture (remember trucker hats?) was widespread. If Pabst couldn't get people to drink its watery brew sincerely, Stewart figured, maybe they could get people to drink it ironically. Pabst began to sponsor hipster events—gallery openings, bike messenger races, snowboarding competitions, and the like. Within a year, sales were way up—which is why, if you walk into a bar in certain Brooklyn neighborhoods, Pabst is more likely to be available than other low-end American beers.

That's not the only excursion in reinvention that Pabst did. In China, where it is branded a "world-famous spirit," Pabst has made itself into a luxury beverage for the cosmopolitan elite. Advertisements compare it to "Scotch whisky, French brandy, Bordeaux wine," and present it in a fluted champagne glass atop a wooden cask. A bottle runs about $44 in U.S. currency.

What's interesting about the Pabst story is that it's not rebranding of the typical sort, in which a product aimed at one group is "repositioned" to appeal to another. Plenty of white working-class men still drink Pabst sincerely, an affirmation of down-home culture. Urban hipsters drink it with a wink. And wealthy Chinese yuppies drink it as a champagne substitute and a signifier of conspicuous consumption. The same beverage means very different things to different people.

Driven by the centrifugal pull of different market segments—each of which wants products that represent its identity—political leadership is fragmenting in much the same way as PBR. Much has been made of Barack Obama's chameleonic political style. "I serve as a blank screen," he wrote in *The Audacity of Hope* in 2006, "on which people of vastly different political stripes project their own views." Part of that is a result of Obama's intrinsic political versatility. But it's also a plus in an age of fragmentation.

(To be sure, the Internet can also facilitate consolidation, as Obama learned when his comment about people "clinging to guns and religion" to donors in San Francisco was reported by the *Huffington Post* and became a top campaign talking point against him. At the same time, Williamsburg hipsters who read the right blogs can learn about Pabst's Chinese marketing scheme. But while this makes fragmentation a more perilous process and cuts against authenticity, it doesn't fundamentally change the calculus. It just makes it more of an imperative to target well.)

The downside of this fragmentation, as Obama has learned, is that it is harder to lead. Acting different with different political constituencies isn't new—in fact, it's probably about as old as politics itself. But the overlap—content that remains constant between all of those constituencies—is shrinking dramatically. You can stand for lots of different kinds of people or stand for something, but doing both is harder every day.

Personalization is both a cause and an effect of the brand fragmentation process. The filter bubble wouldn't be so appealing if it didn't play to our postmaterial desire to maximize self-expression. But once we're in it, the process of matching who we are to content streams can lead to the erosion of common experience, and it can stretch political leadership to the breaking point.

Discourse and Democracy

The good news about postmaterial politics is that as countries become wealthier, they'll likely become more tolerant, and their citizens will be more self-expressive. But there's a dark side to it too. Ted Nordhaus, a student of Inglehart's who focuses on postmaterialism in the environmental movement, told me that "the shadow that comes with postmaterialism is profound self-involvement. . . . We lose all perspective on the collective endeavors that have made the extraordinary lives we live possible." In a postmaterial world where your highest task is to express yourself, the public infrastructure that supports this kind of expression falls out of the picture. But while we can lose sight of our shared problems, they don't lose sight of us.

A few times a year when I was growing up, the nine-hundred-person hamlet of Lincolnville, Maine, held a town meeting. This was my first impression of democracy: A few hundred residents crammed into the grade school auditorium or basement to discuss school additions, speed limits, zoning regulations, and hunting ordinances. In the aisle between the rows of gray metal folding chairs was a microphone on a stand, where people would line up to say their piece.

It was hardly a perfect system: Some speakers droned on; others were shouted down. But it gave all of us a sense of the kinds of people that made up our community that we wouldn't have gotten anywhere else. If the discussion was about encouraging more businesses along the coast, you'd hear from the wealthy summer vacationers who enjoyed their peace and quiet, the back-to-the-land hippies with antidevelopment sentiments, and the families who'd lived in rural poverty for generations and saw the influx as a way up and out. The conversation went back and forth, sometimes closing toward consensus, sometimes fragmenting into debate, but usually resulting in a decision about what to do next.

I always liked how those town meetings worked. But it wasn't until I read *On Dialogue* that I fully understood what they accomplished.

Born to Hungarian and Lithuanian Jewish furniture store owners in Wilkes-Barre, Pennsylvania, David Bohm came from humble roots. But when he arrived at the University of California–Berkeley, he quickly fell in with a small group of theoretical physicists, under the direction of Robert Oppenheimer, who were racing to build the atomic bomb. By the time he died at seventy-two in October 1992, many of his colleagues would remember Bohm as one of the great physicists of the twentieth century.

But if quantum math was his vocation, there was another matter that took up much of Bohm's time. Bohm was preoccupied with the problems created by advanced civilization, especially the possibility of nuclear war. "Technology keeps on advancing with greater and greater power, either for good or for destruction," he wrote. "What is the source of all this trouble? I'm saying that the source is basically in thought." For Bohm, the solution became clear: It was dialogue. In 1992, one of his definitive texts on the subject was published.

To communicate, Bohm wrote, literally means to make something common. And while sometimes this process of making common involves simply sharing a piece of data with a group, more often it involves the group's coming together to create a new, common meaning. "In dialogue," he writes, "people are participants in a pool of common meaning."

Bohm wasn't the first theorist to see the democratic potential of dialogue. Jurgen Habermas, the dean of media theory for much of the twentieth century, had a similar view. For both, dialogue was special because it provided a way for a group of people to democratically create their culture and to calibrate their ideas in the world. In a way, you couldn't have a functioning democracy without it.

Bohm saw an additional reason why dialogue was useful: It provided people with a way of getting a sense of the whole shape of a complex system, even the parts that they didn't directly participate in. Our tendency, Bohm says, is to rip apart and fragment ideas and conversations into bits that have no relation to the whole. He used the example of a watch that has been shattered: Unlike the parts that made up the watch previously, the pieces have no relation to the watch as a whole. They're just little bits of glass and metal.

It's this quality that made the Lincolnville town meetings something special. Even if the group couldn't always agree on where to go, the process helped to develop a shared map for the terrain. The parts understood our relationship to the whole. And that, in turn, makes democratic governance possible.

The town meetings had another benefit: They equipped us to deal more handily with the problems that did emerge. In the science of social mapping, the definition of a community is a set of nodes that are densely interconnected—my friends form a community if they all don't know just me but also have independent relationships with one another. Communication builds stronger community.

Ultimately, democracy works only if we citizens are capable of thinking beyond our narrow self-interest. But to do so, we need a shared view of the world we cohabit. We need to come into contact with other peoples' lives and needs and desires. The filter bubble pushes us in the opposite direction—it creates the impression that our narrow self-interest is all that exists. And while this is great for getting people to shop online, it's not great for getting people to make better decisions together.

"The prime difficulty" of democracy, John Dewey wrote, "is that of discovering the means by which a scattered, mobile, and manifold public may so recognize itself as to define and express its interests." In the early days of the Internet, this was one of the medium's great hopes—that it would finally offer a medium whereby whole towns—and indeed countries—could co-create their culture through discourse. Personalization has given us something very different: a public sphere sorted and manipulated by algorithms, fragmented by design, and hostile to dialogue.

Which begs an important question: Why would the engineers who designed these systems want to build them this way?

Hello, World!

> SOCRATES: Or again, in a ship, if a man having the power to do what he likes, has no intelligence or skill in navigation [αρετης κυβερνητικης, *aretēs kybernētikēs*], do you see what will happen to him and to his fellow-sailors?
>
> —*Plato, First Alcibiades,* the earliest known use of the word *cybernetics*

It's the first fragment of code in the code book, the thing every aspiring programmer learns on day one. In the C++ programming language, it looks like this:

```
void main()
{
cout << "Hello, World!" <<
endl;
}
```

Although the code differs from language to language, the result is the same: a single line of text against a stark white screen:

Hello, World!

A god's greeting to his invention—or perhaps an invention's greeting to its god. The delight you experience is electric—the current of creation, running through your fingers into the keypad, into the machine, and back out into the world. *It's alive!*

That every programmer's career begins with "Hello, World!" is not a coincidence. It's the power to create new universes, which is what often draws people to code in the first place. Type in a few lines, or a few thousand, strike a key, and something seems to come to life on your screen—a new space unfolds, a new engine roars. If you're clever enough, you can make and manipulate anything you can imagine.

"We are as Gods," wrote futurist Stewart Brand on the cover of his *Whole Earth Catalog* in 1968, "and we might as well get good at it." Brand's catalog, which sprang out of the back-to-the-land movement, was a favorite among California's emerging class of programmers and computer enthusiasts. In Brand's view, tools and technologies turned people, normally at the mercy of their environments, into gods in control of them. And the computer was a tool that could become any tool at all.

Brand's impact on the culture of Silicon Valley and geekdom is hard to overestimate—though he wasn't a programmer himself, his vision shaped the Silicon Valley worldview. As Fred Turner details in the fascinating *From Counterculture to Cyberculture*, Brand and his cadre of do-it-yourself futurists were disaffected hippies—social revolutionaries who were uncomfortable with the communes sprouting up in Haight-Ashbury. Rather than seeking to build a new world through political change, which required wading through the messiness of compromise and group decision making, they set out to build a world on their own.

In *Hackers*, his groundbreaking history of the rise of engineering culture, Steve Levy points out that this ideal spread from the programmers themselves to the users "each time some user flicked the machine on, and the screen came alive with words, thoughts, pictures, and sometimes elaborate worlds built out of air—those computer programs which could make any man (or woman) a god." (In the era described by Levy's book, the term *hacker* didn't have the transgressive, law-breaking connotations it acquired later.)

The God impulse is at the root of many creative professions: Artists conjure up color-flecked landscapes, novelists build whole societies on paper. But it's always clear that these are creations: A painting doesn't talk back. A program can, and the illusion that it's "real" is powerful. Eliza, one of the first rudimentary AI programs, was programmed with a battery of therapistlike questions and some basic contextual cues. But students spent hours talking to it about their deepest problems: "I'm having some troubles with my family," a student might write, and Eliza would immediately respond, "Tell me more about your family."

Especially for people who've been socially ostracized due to quirks or brains or both, there are at least two strong draws to the world-building impulse. When social life is miserable or oppressive, escapism is a reasonable response—it's probably not coincidental that role-playing games, sci-fi and fantasy literature, and programming often go together.

The infinitely expandable universe of code provides a second benefit: complete power over your domain. "We all fantasize about living without rules," says Siva Vaidyanathan. "We imagine the Adam Sandler movie where you can move around and take people's clothes off. If you don't think of reciprocity as one of the beautiful and rewarding things about being a human being, you wish for a place or a way of acting without consequence." When the rules of high school social life seem arbitrary and oppressive, the allure of making your own rules is pretty powerful.

This approach works pretty well as long as you're the sole denizen of your creation. But like the God of Genesis, coders quickly get lonely. They build portals into their homespun worlds, allowing others to enter. And that's where things get complicated: On the one hand, the more inhabitants in the world you've built, the more power you have. But on the other hand, the citizens can get uppity. "The programmer wants to set up some rules, to either a game or a system, and then let it run without interference from anything," says Douglas Rushkoff, an early cyberbooster-turned-cyberpragmatist. "If you have a program that needs a minder to come in and help it run, then it's not a very good program, is it? It's supposed to just run."

Coders sometimes harbor God impulses; they sometimes even have aspirations to revolutionize society. But they almost never aspire to be politicians. "While programming is considered a transparent, neutral, highly controllable realm . . . where production results in immediate gratification and something useful," writes NYU anthropologist Gabriella Coleman, "politics tends to be seen by programmers as buggy, mediated, tainted action clouded by ideology that is not productive of much of anything." There's some merit to that view, of course. But for programmers to shun politics completely is a problem— because increasingly, given the disputes that inevitably arise when people come together, the most powerful ones will be required to adjudicate and to govern.

Before we get to how this blind spot affects our lives, though, it's worth looking at how engineers think.

The Empire of Clever

Imagine that you're a smart high school student on the low end of the social totem pole. You're alienated from adult authority, but unlike many teenagers, you're also alienated from the power structures of your peers—an existence that can feel lonely and peripheral. Systems and equations are intuitive, but people aren't—social signals are confusing and messy, difficult to interpret.

Then you discover code. You may be powerless at the lunch table, but code gives you power over an infinitely malleable world and opens the door to a symbolic system that's perfectly clear and ordered. The jostling for position and status fades away. The nagging parental voices disappear. There's just a clean, white page for you to fill, an opportunity to build a better place, a home, from the ground up.

No wonder you're a geek.

This isn't to say that geeks and software engineers are friendless or even socially inept. But there's an implicit promise in becoming a coder: Apprentice yourself to symbolic systems, learn to carefully understand the rules that govern them, and you'll gain power to manipulate them. The more powerless you feel, the more appealing this promise becomes. "Hacking," Steven Levy writes, "gave you not only an understanding of the system but an addictive control as well, along with the illusion that total control was just a few features away."

As anthropologist Coleman points out, beyond the Jocks-and-Nerds stereotypes, there are actually many different geek cultures. There are open-software advocates, most famously embodied by Linux founder Linus Torvalds, who spend untold hours collaboratively building free software tools for the masses, and there are Silicon Valley start-up entrepreneurs. There are antispam zealots, who organize online posses to seek out and shut down Viagra purveyors. And then there's the more antago-nistic wing: spammers; "trolls," who spend their time looking for fun ways to leverage technology at others' expense; "phreaks," who are animated by the challenge to break open telecommu-nications systems; and hackers who break into government systems to prove it can be done.

Generalizations that span these different niches and com-munities run the risk of stereotyping and tend to fall short. But at the heart of these subcultures is a shared method for looking at and asserting power in the world, which influences how and why online software is made.

The through-line is a focus on systematization. Nearly all geek cultures are structured as an empire of clever wherein ingenuity, not charisma, is king. The intrinsic efficiency of a cre-ation is more important than how it looks. Geek cultures are data driven and reality based, valuing substance over style. Humor plays a prominent role—as Coleman points out, jokes demonstrate an ability to manipulate language in the same way that an elegant solution to a tricky programming problem demonstrates mastery over code. (The fact that humor also often serves to unmask the ridiculous pieties of the powerful is undoubtedly also part of its appeal.)

Systematization is especially alluring because it doesn't offer power just in the virtual sphere. It can also provide a way to understand and navigate social situations. I learned this firsthand when, as an awkward seventeen-year-old with all the trappings of geek experience (the fantasy books, the introversion, the obsession with HTML and BBSes), I flew across the country to accept the wrong job.

In a late-junior-year panic, I'd applied for every internship I could find. One group, a nuclear disarmament organization based in San Francisco, had gotten back to me, and without much further investigation, I'd signed up. It was only when I walked into the office that I realized I'd signed up to be a canvaser. Off the top of my head, I couldn't imagine a worse fit, but because I had no other prospects, I decided to stick out the day of training.

Canvasing, the trainer explained, was a science as much as an art. And the laws were powerful. Make eye contact. Explain why the issue matters to you. And after you ask for money, let your target say the first thing. I was intrigued: Asking people for money was scary, but the briefing hinted at a hidden logic. I committed the rules to memory.

When I walked through my first grassy Palo Alto lawn, my

heart was in my throat. Here I was at the doorstep of someone I'd never met, asking for $50. The door opened and a harried woman with long gray hair peeped out. I took a deep breath, and launched into my spiel. I asked. I waited. And then she nodded and went to get her checkbook.

The euphoria I felt wasn't about the $50. It was about something bigger—the promise that the chaos of human social life could be reduced to rules that I could understand, follow, and master. Conversation with strangers had never come naturally to me—I didn't know what to talk about. But the hidden logic of getting someone I'd never met to trust me with $50 had to be the tip of a larger iceberg. By the end of a summer traipsing through the yards of Palo Alto and Marin, I was a master canvaser.

Systematization is a great method for building functional software. And the quantitative, scientific approach to social observation has given us many great insights into human phenomena as well. Dan Ariely researches the "predictably irrational" decisions we make on a daily basis; his findings help us make better decisions. The blog at OkCupid.com, a math-driven dating Web site, identifies patterns in the e-mails flying back and forth between people to make them better daters ("Howdy" is a better opener than "Hi").

But there are dangers in taking the method too far. As I discussed in chapter 5, the most human acts are often the most unpredictable ones. Because systematizing works much of the time, it's easy to believe that by reducing and brute-forcing an understanding of any system, you can control it. And as a master of a self-created universe, it's easy to start to view people as a means to an end, as variables to be manipulated on a mental spreadsheet, rather than as breathing, thinking beings. It's difficult both to systematize and to appeal to the fullness of human life—its unpredictability, emotionality, and surprising quirks—at the same time.

David Gelernter, a Yale computer scientist, barely survived an encounter with an explosive package sent by the Unabomber; his eyesight and right hand are permanently damaged as a result. But Gelernter is hardly the technological utopian Ted Kaczinski believed him to be.

"When you do something in the public sphere," Gelernter told a reporter, "it behooves you to know something about what the public sphere is like. How did this country get this way? What was the history of the relationship between technology and the public? What's the history of political exchange? The problem is, hackers don't tend to know any of that. And that's why it worries me to have these people in charge of public policy. Not because they're bad, just because they're uneducated."

Understanding the rules that govern a messy, complex world makes it intelligible and navigable. But systematizing inevitably involves a trade-off—rules give you some control, but you lose nuance and texture, a sense of deeper connection. And when a strict systematizing sensibility entirely shapes social space (as it often does online), the results aren't always pretty.

The New Architects

The political power of design has long been obvious to urban planners. If you take the Wantagh State Parkway from

Westbury to Jones Beach on Long Island, at intervals you'll pass under several low, vine-covered overpasses. Some of them have as little as nine feet of clearance. Trucks aren't allowed on the parkway—they wouldn't fit. This may seem like a design oversight, but it's not.

There are about two hundred of these low bridges, part of the grand design for the New York region pioneered by Robert Moses. Moses was a master deal maker, a friend of the great politicians of the time, and an unabashed elitist. According to his biographer, Robert A. Caro, Moses's vision for Jones Beach was as an island getaway for middle-class white families. He included the low bridges to make it harder for low-income (and mostly black) New Yorkers to get to the beach, as public buses—the most common form of transport for inner-city residents—couldn't clear the overpasses.

The passage in Caro's *The Power Broker* describing this logic caught the eye of Langdon Winner, a *Rolling Stone* reporter, musician, professor, and philosopher of technology. In a pivotal 1980 article titled "Do Artifacts Have Politics?" Winner considered how Moses's "monumental structures of concrete and steel embody a systematic social inequality, a way of engineering relationships among people that, after a time, became just another part of the landscape."

On the face of it, a bridge is just a bridge. But often, as Winner points out, architectural and design decisions are underpinned by politics as much as aesthetics. Like goldfish that grow only large enough for the tank they're in, we're contextual beings: how we behave is dictated in part by the shape of our environments. Put a playground in a park, and you encourage one kind of use; build a memorial, and you encourage another.

As we spend more of our time in cyberspace—and less of our time in what geeks sometimes call meatspace, or the offline world—Moses's bridges are worth keeping in mind. The algorithms of Google and Facebook may not be made of steel and concrete, but they regulate our behavior just as effectively. That's what Larry Lessig, a law professor and one of the early theorists of cyberspace, meant when he famously wrote that "code is law."

If code is law, software engineers and geeks are the ones who get to write it. And it's a funny kind of law, created without any judicial system or legislators and enforced nearly perfectly and instantly. Even with antivandalism laws on the books, in the physical world you can still throw a rock through the window of a store you don't like. You might even get away with it. But if vandalism isn't part of the design of an online world, it's simply impossible. Try to throw a rock through a virtual storefront, and you just get an error.

Back in 1980, Winner wrote, "Consciously or unconsciously, deliberately or inadvertently, societies choose structures for technologies that influence how people are going to work, communicate, travel, consume, and so forth over a very long time." This isn't to say that today's designers have malevolent impulses, of course—or even that they're always explicitly trying to shape society in certain ways. It's just to say that they can—in fact, they can't help but shape the worlds they build.

To paraphrase Spider-Man creator Stan Lee, with great power comes great responsibility. But the programmers who brought us the Internet and now the filter bubble aren't always game to take on that responsibility. The Hacker Jargon File, an online repository of geek culture, puts it this way: "Hackers are far more likely than most non-hackers to either (a) be aggressively apolitical or (b) entertain peculiar or idiosyncratic political ideas." Too often, the executives of Facebook, Google, and other socially important companies play it coy: They're social revolutionaries when it suits them and neutral, amoral businessmen when it doesn't. And both approaches fall short in important ways.

Playing It Coy

When I first called Google's PR department, I explained that I wanted to know how Google thought about its enormous curatorial power. What was the code of ethics, I asked, that Google uses to determine what to show to whom? The public affairs manager on the other end of the phone sounded confused. "You mean privacy?" No, I said, I wanted to know how Google thought about its editorial power. "Oh," he replied, "we're just trying to give people the most relevant information." Indeed, he seemed to imply, no ethics were involved or required.

I persisted: If a 9/11 conspiracy theorist searches for "9/11," was it Google's job to show him the *Popular Mechanics* article that debunks his theory or the movie that supports it? Which was more relevant? "I see what you're getting at," he said. "It's an interesting question." But I never got a clear answer.

Much of the time, as the Jargon File entry claims, engineers

resist the idea that their work has moral or political consequences at all. Many engineers see themselves as interested in
efficiency and design, in building cool stuff rather than messy
ideological disputes and inchoate values. And it's true that if
political consequences of, say, a somewhat faster video-rendering
engine exist, they're pretty obscure.

But at times, this attitude can verge on a "Guns don't kill
people, people do" mentality—a willful blindness to how their
design decisions affect the daily lives of millions. That Facebook's button is named Like prioritizes some kinds of information over others. That Google has moved from PageRank—which
is designed to show the societal consensus result—to a mix of
PageRank and personalization represents a shift in how Google
understands relevance and meaning.

This amorality would be par for the corporate course if it
didn't coincide with sweeping, world-changing rhetoric from
the same people and entities. Google's mission to organize the
world's information and make it accessible to everyone carries
a clear moral and even political connotation—a democratic
redistribution of knowledge from closed-door elites to the people. Apple's devices are marketed with the rhetoric of social
change and the promise that they'll revolutionize not only your
life but our society as well. (The famous Super Bowl ad an-
nouncing the release of the Macintosh computer ends by
declaring that "1984 won't be like *1984*.")

Facebook describes itself as a "social utility," as if it's a
twenty-first-century phone company. But when users protest
Facebook's constantly shifting and eroding privacy policy,
Zuckerberg often shrugs it off with the caveat emptor posture

that if you don't want to use Facebook, you don't have to. It's hard to imagine a major phone company getting away with saying, "We're going to publish your phone conversations for anyone to hear—and if you don't like it, just don't use the phone."

Google tends to be more explicitly moral in its public aspirations; its motto is "Don't be evil," while Facebook's unofficial motto is "Don't be lame." Nevertheless, Google's founders also sometimes play a get-out-of-jail-free card. "Some say Google is God. Others say Google is Satan," says Sergey Brin. "But if they think Google is too powerful, remember that with search engines, unlike other companies, all it takes is a single click to go to another search engine. People come to Google because they choose to. We don't trick them."

Of course, Brin has a point: No one is forced to use Google, just as no one is forced to eat at McDonald's. But there's also something troubling about this argument, which minimizes the responsibility he might have to the billions of users who rely on the service Google provides and in turn drive the company's billions in advertising revenue.

To further muddle the picture, when the social repercussions of their work are troubling, the chief architects of the online world often fall back on the manifest-destiny rhetoric of technodeterminism. Technologists, Siva Vaidyanathan points out, rarely say something "could" or "should" happen—they say it "will" happen. "The search engines of the future will be personalized," says Google Vice President Marissa Mayer, using the passive tense.

Just as some Marxists believed that the economic conditions of a society would inevitably propel it through capitalism and toward a world socialist regime, it's easy to find engineers and technodeterminist pundits who believe that technology is on a set course. Sean Parker, the cofounder of Napster and rogue early president of Facebook, tells *Vanity Fair* that he's drawn to hacking because it's about "re-architecting society. It's technology, not business or government, that's the real driving force behind large-scale societal shifts."

Kevin Kelly, the founding editor of *Wired*, wrote perhaps the boldest book articulating the technodeterminist view, *What Technology Wants*, in which he posits that technology is a "seventh kingdom of life," a kind of meta-organism with desires and tendencies of its own. Kelly believes that the technium, as he calls it, is more powerful than any of us mere humans. Ultimately, technology—a force that "wants" to eat power and expand choice—will get what it wants whether we want it to or not.

Technodeterminism is alluring and convenient for newly powerful entrepreneurs because it absolves them of responsibility for what they do. Like priests at the altar, they're mere vessels of a much larger force that it would be futile to resist. They need not concern themselves with the effects of the systems they've created. But technology *doesn't* solve every problem of its own accord. If it did, we wouldn't have millions of people starving to death in a world with an oversupply of food.

It shouldn't be surprising that software entrepreneurs are incoherent about their social and political responsibilities. A great deal of this tension undoubtedly comes from the fact that the nature of online business is to scale up as quickly as possible. Once you're on the road to mass success and riches—often as a very young coder—there simply isn't much time to fully think all of this through. And the pressure of the venture capitalists breathing down your neck to "monetize" doesn't always offer much space for rumination on social responsibility.

The $50 Billion Sand Castle

Once a year, the Y Combinator start-up incubator hosts a day-long conference called Startup School, where successful tech entrepreneurs pass wisdom on to the aspiring audience of bright-eyed Y Combinator investees. The agenda typically includes many of the top CEOs in Silicon Valley, and in 2010, Mark Zuckerberg was at the top of the list.

Zuckerberg was in an affable mood, dressed in a black T-shirt and jeans and enjoying what was clearly a friendly crowd. Even so, when Jessica Livingston, his interviewer, asked him about *The Social Network*, the movie that had made him a household name, a range of emotions crossed his face. "It's interesting what kind of stuff they focused on getting right," Zuckerberg began. "Like, every single shirt and fleece they had in that movie is actually a shirt or fleece that I own."

Where there was an egregious discrepancy between fiction and reality, Zuckerberg told her, was how his own motivations were painted. "They frame it as if the whole reason for making Facebook and building something was that I wanted to get girls, or wanted to get into some kind of social institution. And the reality, for people who know me, is that I've been dating the same girl since before I started Facebook. It's such a big disconnect. . . . They just can't wrap their head around the idea that someone might build something because they like building things."

It's entirely possible that the line was just a clever bit of Facebook PR. And there's no question that the twenty-six-year-old billionaire is motivated by empire building. But the comment struck me as candid: For programmers as for artists and craftsmen, making things is often its own best reward.

Facebook's flaws and its founder's ill-conceived views about identity aren't the result of an antisocial, vindictive mind-set. More likely, they're a natural consequence of the odd situation successful start-ups like Facebook create, in which a twenty-something guy finds himself, in a matter of five years, in a position of great authority over the doings of 500 million human beings. One day you're making sand castles; the next, your sand castle is worth $50 billion and everyone in the world wants a piece of it.

Of course, there are far worse business-world personality types with whom to entrust the fabric of our social lives. With a reverence for rules, geeks tend to be principled—to carefully consider and then follow the rules they set for themselves and to stick to them under social pressure. "They have a somewhat skeptical view of authority," Stanford professor Terry Winograd said of his former students Page and Brin. "If they see the world going one way and they believe it should be going the other way, they are more like to say 'the rest of the world is wrong' rather than 'maybe we should reconsider.'"

But the traits that fuel the best start-ups—aggression, a touch of arrogance, an interest in empire building, and of course brilliant systematizing skills—can become a bit more problematic when you rule the world. Like pop stars who are vaulted onto the global stage, world-building engineers aren't always ready or willing to accept the enormous responsibility they come to hold when their creations start to teem with life. And it's not infrequently the case that engineers who are deeply mistrustful of power in the hands of others see themselves as supreme rationalists impervious to its effects.

It may be that this is too much power to entrust to any small, homogeneous group of individuals. Media moguls who get their start with a fierce commitment to the truth become the confidants of presidents and lose their edge; businesses begun as social ventures become preoccupied with delivering shareholder value. In any case, one consequence of the current system is that we can end up placing a great deal of power in the hands of people who can have some pretty far-out, not entirely well-developed political ideas. Take Peter Thiel, one of Zuckerberg's early investors and mentors.

Thiel has penthouse apartments in San Francisco and New York and a silver gullwing McLaren, the fastest car in the world. He also owns about 5 percent of Facebook. Despite his boyish, handsome features, Thiel often looks as though he's brooding. Or maybe he's just lost in thought. In his teenage years, he was a high-ranking chess player but stopped short of becoming a grand master. "Taken too far, chess can become an alternate reality in which one loses sight of the real world," he told an interviewer for *Fortune*. "My chess ability was roughly at the limit. Had I become any stronger, there would have been some massive tradeoffs with success in other domains in life." In high school, he read Solzhenitsyn's *Gulag Archipelago* and J. R. R. Tolkien's *Lord of the Rings* series, visions of corrupt and totalitarian power. At Stanford, he started a libertarian newspaper, the *Stanford*, to preach the gospel of freedom.

In 1998, Thiel cofounded the company that would become PayPal, which he sold to eBay for $1.5 billion in 2002. Today Thiel runs a multi-billion-dollar hedge fund, Clarium, and a venture capital firm, Founder's Fund, which invests in software companies throughout Silicon Valley. Thiel has made some legendarily good picks—among them, Facebook, in which he was the first outside investor. (He's also made some bad ones—Clarium has lost billions in the last few years.) But for Thiel, investing is more than a day job. It's an avocation. "By starting a new Internet business, an entrepreneur may create a new world," Thiel says. "The hope of the Internet is that these new worlds will impact and force change on the existing social and political order."

His comments raise the question of what kind of change Thiel would like to see. While many billionaires are fairly circumspect about their political views, Thiel has been vocal— and it's safe to say that there are few with views as unusual as Thiel's. "Peter wants to end the inevitability of death and taxes," Thiel's sometime collaborator Patri Friedman (grandson of Milton) told *Wired*. "I mean, talk about aiming high!"

In an essay posted on the libertarian Cato Institute's Web site, Thiel describes why he believes that "freedom and democracy are no longer compatible." "Since 1920," he writes, "the vast increase in welfare beneficiaries and the extension of the franchise to women—two constituencies that are notoriously tough for libertarians—have rendered the notion of 'capitalist democracy' into an oxymoron." Then he outlines his hopes for the future: space exploration, "sea-steading," which involves building movable microcountries on the open ocean, and cyberspace. Thiel has poured millions into technologies to sequence genes and prolong life. He's also focused on preparing for the Singularity, the moment a few decades from now when some futurists believe that humans and machines are likely to meld.

In an interview, he argues that should the Singularity arrive, one would be well advised to be on the side of the computers: "Certainly we would hope that [an artificially intelligent computer] would be friendly to human beings. At the same time, I don't think you'd want to be known as one of the human beings that is against computers and makes a living being against computers."

If all this sounds a little fantastical, it doesn't worry Thiel. He's focused on the long view. "Technology is at the center of what will determine the course of the 21st century," he says. "There are aspects of it that are great and aspects that are terrible, and there are some real choices humans have to make about which technologies to foster and which ones we should be more careful about."

Peter Thiel is entitled to his idiosyncratic views, of course, but they're worth paying attention to because they increasingly shape the world we all live in. There are only four other people on the Facebook board besides Mark Zuckerberg; Thiel is one of them, and Zuckerberg publicly describes him as a mentor. "He helped shape the way I think about the business," Zuckerberg said in a 2006 Bloomberg News interview. As Thiel says, we have some big decisions to make about technology. And as for how those decisions get made? "I have little hope," he writes, "that voting will make things better."

"What Game Are You Playing?"

Of course, not all engineers and geeks have the views about democracy and freedom that Peter Thiel does—he's surely an outlier. Craig Newmark, the founder of the free Web site craigslist, spends most of his time arguing for "geek values" that include service and public-spiritedness. Jimmy Wales and the editors at Wikipedia work to make human knowledge free to everyone. The filtering goliaths make huge contributions here as well: The democratic ideal of an enlightened, capable citizenry is well served by the broader set of relationships Facebook allows me to manage and the mountains of formerly hard-to-access research papers and other public information that Google has freed.

But the engineering community can do more to strengthen the Internet's civic space. And to get a sense of the path ahead, I talked to Scott Heiferman.

Heiferman, the founder of MeetUp.com, is soft-spoken in a Midwestern sort of way. That's fitting, because he grew up in Homewood, Illinois, a small town on the outskirts of Chicago. "It was a stretch to call it suburban," he says. His parents operated a paint store.

As a teenager, Heiferman devoured material about Steve Jobs, eating up the story about how Jobs wooed a senior executive from Pepsi by asking him if he wanted to change the world or sell sugar water. "Throughout my life," he told me, "I've had a love-hate relationship with advertising." At the University of Iowa in the early 1990s, Heiferman studied engineering and marketing but at night he ran a radio show called *Advertorial*

Infotainment in which he would remix and cut advertisements together to create a kind of sound art. He put the finished shows online and encouraged people to send in ads to remix, and the attention got him his first job, managing the Web site at Sony .com.

After a few years as Sony's "interactive-marketing frontiersman," Heiferman founded i-traffic, one of the major early advertising companies of the Web. Soon i-traffic was the agency of record for clients like Disney and British Airways. But although the company was growing quickly, he was dissatisfied. The back of his business card had a mission statement about connecting people with brands they'd love, but he was increasingly uncertain that was a worthy endeavor—perhaps he was selling sugar water after all. He left the company in 2000.

For the remainder of the year and into 2001, Heiferman was in a funk. "I was exhibiting what you could call being depressed," he says. When he heard the first word of the World Trade Center attacks on 9/11, he ran up to his lower-Manhattan rooftop and watched in horror. "I talked to more strangers in the next three days," he says, "than in the previous five years of living in New York."

Shortly after the attacks, Heiferman came across the blog post that changed his life. It argued that the attacks, as awful as they were, might bring Americans back together in their civic life, and referenced the bestselling book *Bowling Alone*. Heiferman bought a copy and read it cover to cover. "I became captivated," he says, "by the question of whether we could use technology to rebuild and strengthen community." MeetUp.com, a site that makes it easy for local groups to meet face-to-face, was his answer, and today, MeetUp serves over 79,000 local groups that do that. There's the Martial Arts MeetUp in Orlando and the Urban Spirituality MeetUp in Barcelona and the Black Singles MeetUp in Houston. And Heiferman is a happier man.

"What I learned being in the ad business," he says, "is that people can just go a long time without asking themselves what they should put their talent towards. You're playing a game, and you know the point of the game is to win. But what game are you playing? What are you optimizing for? If you're playing the game of trying to get the maximum downloads of your app, you'll make the better farting app."

"We don't need more things," he says. "People are more magical than iPads! Your relationships are not media. Your friendships are not media. Love is not media." In his low-key way, Heiferman is getting worked up.

Evangelizing this view of technology—that it ought to do something meaningful to make our lives more fulfilling and to solve the big problems we face—isn't as easy as it might seem. In addition to MeetUp more generally, Scott founded the New York Tech MeetUp, a group of ten thousand software engineers who meet every month to preview new Web sites. At a recent meeting, Scott made an impassioned plea for the assembled group to focus on solving the problems that matter—education, health care, the environment. It didn't get a very good reception—in fact, he was just about booed off the stage. "'We just want to do cool stuff,' was the attitude," Scott told me later. "'Don't bother me with this politics stuff.'"

Technodeterminists like to suggest that technology is inherently good. But despite what Kevin Kelly says, technology is no more benevolent than a wrench or a screwdriver. It's only good when people make it do good things and use it in good ways.

Melvin Kranzberg, a professor who studies the history of technology, put it best nearly thirty years ago, and his statement is now known as Kranzberg's first law: "Technology is neither good or bad, nor is it neutral."

For better or worse, programmers and engineers are in a position of remarkable power to shape the future of our society. They can use this power to help solve the big problems of our age—poverty, education, disease—or they can, as Heiferman says, make a better farting app. They're entitled to do either, of course. But it's disingenuous to have it both ways—to claim your enterprise is great and good when it suits you and claim you're a mere sugar-water salesman when it doesn't.

Actually, building an informed and engaged citizenry—in which people have the tools to help manage not only their own lives but their own communities and societies—is one of the most fascinating and important engineering challenges. Solving it will take a great deal of technical skill mixed with humanistic understanding—a real feat. We need more programmers to go beyond Google's famous slogan, "Don't be evil." We need engineers who will do good.

And we need them soon: If personalization remains on its current trajectory, as the next chapter describes, the near future could be stranger and more problematic than many of us would imagine.

What You Want, Whether You Want It or Not

> There will always be plenty of things to compute in the detailed affairs of millions of people doing complicated things.
>
> —computing pioneer Vannevar Bush, 1945

> All collected data had come to a final end. Nothing was left to be collected. But all collected data had yet to be completely correlated and put together in all possible relationships.
>
> —from Isaac Asimov's short story "The Last Question"

I recently received a friend invitation on Facebook from someone whose name I didn't recognize, a curvy-figured girl with big eyes and thick lashes. Clicking to figure out who she was (and, I'll admit, to look more closely), I read over her profile. It didn't tell me a lot about her, but it seemed like the profile of someone I might plausibly know. A few of our interests were the same.

I looked again at the eyes. They were a little *too* big.

In fact, when I looked more closely, I realized her profile picture wasn't even a photograph—it had been rendered by a 3-D graphics program. There was no such person. My new attractive would-be friend was a figment of software, crawling through friend connections to harvest data from Facebook users. Even the list of movies and books she liked appeared to have been ripped from the lists of her "friends."

For lack of a better word, let's call her an *advertar*—a virtual being with a commercial purpose. As the filter bubble's membrane becomes thicker and harder to penetrate, advertars could become a powerful adaptive strategy. If I only get the news from my code and my friends, the easiest way to get my attention might be friends who are code.

The technologies that support personalization will only get more powerful in the years ahead. Sensors that can pick up new personal signals and data streams will become even more deeply embedded in the surface of everyday life. The server farms that support the Googles and Amazons will grow, while the processors inside them shrink; that computing power will be unleashed to make increasingly precise guesses about our preferences and even our interior lives. Personalized "augmented reality" technologies will project an overlay over our experience of the real world, not just the digital one. Even Nicholas Negroponte's intelligent agents may make a comeback. "Markets are strong forces," says Bill Joy, the legendary programmer who cofounded Sun Microsystems. "They take you somewhere very quickly. And if where they take you is not where you want to go, you've got a problem."

In 2002, the sci-fi movie *Minority Report* featured personalized holographic advertisements that accosted pedestrians as they walked down the street. In Tokyo, the first *Minority Report*–style personalized billboard has gone up outside of the NEC corporation's headquarters (minus, for now, the holography). It's powered by the company's PanelDirector software, which scans the faces of passersby and matches them to a database of ten thousand stored photos to make guesses about their age and gender. When a young woman steps in front of the display, it responds instantly by showing her ads tailored to her. IBM's on the case, too; its prototype advertising displays use remotely readable identity cards to greet viewers by name.

In *Reality Hunger*, a book-length essay composed entirely of text fragments and reworked quotations, David Shields makes the case for the growing movement of artists who are "breaking larger and larger chunks of 'reality' into their work." Shields's examples are far-ranging, including *The Blair Witch Project*, *Borat*, and *Curb Your Enthusiasm*; karaoke, VH1's *Behind the Music*, and public access TV; *The Eminem Show* and *The Daily Show*, documentary and mockumentary. These pieces, he says, are the most vital art of our time, part of a new mode characterized by "a deliberate unartiness" and "a blurring (to the point of invisibility) of any distinction between fiction and nonfiction: the lure and blur of the real." Truthiness, in Shields's view, is the future of art.

As goes art, so goes technology. The future of personalization—and of computing itself—is a strange amalgam of the real and the virtual. It's a future where our cities and our bedrooms and all of the spaces in between exhibit what researchers call "ambient intelligence." It's a future where our environments shift around us to suit our preferences and even our moods. And it's a future where advertisers will develop ever more powerful and reality-bending ways to make sure their products are seen.

The days when the filter bubble disappears when we step away from our computers, in other words, are numbered.

The Robot with Gaydar

Stanford Law professor Ryan Calo thinks a lot about robots, but he doesn't spend much time musing about a future of cyborgs and androids. He's more interested in Roombas, the little robotic vacuum cleaners currently on the market. Roomba owners name their machines like pets. They delight in watching the little bumbling devices wander around the room. Roombas provoke an emotional response, even a sense of relationship. And in the next few years, they'll be joined by a small army of consumer-electronic brethren.

The increasing prevalence of humanlike machines in everyday life presents us with new dilemmas in personalization and privacy. The emotions provoked by "humanness," both virtually (advertars) and in reality (humanlike robots) are powerful. And when people begin to relate to machines as we do to humans, we can be convinced to reveal implicit information that we would never directly give away.

For one thing, the presence of humanoid faces changes behavior, compelling people to behave more like they're in public. The Chinese experiment with Jingjing and Chacha, the cartoon Internet police, is one example of this power. On the one hand, Calo points out, people are much less likely to volunteer private information when being interrogated by a virtual agent than when simply filling out a form. This is part of why the intelligent-agent craze didn't work out the first time around: In many cases, it's easier to get people to share personal information if they feel as though they're privately entering it into an impersonal machine rather than sharing it with people.

On the other hand, when Harvard researchers Terence Burnham and Brian Hare asked volunteers to play a game in which they could choose to donate money or keep it, a picture of the friendly looking robot Kismet increased donations by 30 percent. Humanlike agents tend to make us clam up on the intimate details of our lives, because they make us feel as if we're actually around other people. For elderly folks living alone or a child recovering in a hospital, a virtual or robotic friend can be a great relief from loneliness and boredom.

This is all to the good. But humanlike agents also have a great deal of power to shape our behavior. "Computers programmed to be polite, or to evidence certain personalities," Calo writes, "have profound effects on the politeness, acceptance, and other behavior of test subjects." And because they engage with people, they can pull out implicit information that we'd never intend to divulge. A flirty robot, for example, might be able to read subconscious cues—eye contact, body language—to quickly identify personality traits of its interlocutor.

The challenge, Calo says, is that it's hard to remember that humanlike software and hardware aren't human at all. Advertars or robotic assistants may have access to the whole set of personal data that exists online—they may know more about you, more precisely, than your best friend. And as persuasion and personality profiling get better, they'll develop an increasingly nuanced sense of how to shift your behaviors.

Which brings us back to the advertar. In an attention-limited world, lifelike, and especially humanlike, signals stand out—we're hardwired to pay attention to them. It's far easier to ignore a billboard than an attractive person calling your name. And as a result, advertisers may well decide to invest in technology that allows them to insert human advertisements into social spaces. The next attractive man or woman who friends you on Facebook could turn out to be an ad for a bag of chips.

As Calo puts it, "people are not evolved to twentieth-century technology. The human brain evolved in a world in which only humans exhibited rich social behaviors, and a world in which all perceived objects were real physical objects." Now all that's shifting.

The Future Is Already Here

The future of personalization is driven by a simple economic calculation. Signals about our personal behavior and the computing power necessary to crunch through them are becoming cheaper than ever to acquire. And as that cost collapses, strange new possibilities come within reach.

Take facial recognition. Using MORIS, a $3,000 iPhone

app, the police in Brockton, Massachusetts, can snap a photo of a suspect and check his or her identity and criminal record in seconds. Tag a few pictures with Picasa, Google's photo-management tool, and the software can already pick out who's who in a collection of photos. And according to Eric Schmidt, the same is true of Google's cache of images from the entire Web. "Give us 14 images of you," he told a crowd of technologists at the Techonomy Conference in 2010, "and we can find other images of you with ninety-five percent accuracy."

As of the end of 2010, however, this feature isn't available in Google Image Search. Face.com, an Israeli start-up, may offer the service before the search giant does. It's not every day that a company develops a highly useful and world-changing technology and then waits for a competitor to launch it first. But Google has good reason to be concerned: The ability to search by face will shatter many of our cultural illusions about privacy and anonymity.

Many of us will be caught in flagrante delicto. It's not just that your friends (and enemies) will be able to easily find pictures other people have taken of you—as if the whole Internet has been tagged on Facebook. They will also be able to find pictures other people took of other people, in which you happen to be walking by or smoking a cigarette in the background.

After the data has been crunched, the rest is easy. Want to search for two people—say your boyfriend and that overly friendly intern you suspect him of dallying with, or your employee and that executive who's been trying to woo him away? Easy. Want to build a Facebook-style social graph by looking at who appears most often with whom? A cinch. Want to see which of your coworkers posted profiles on anonymous dating sites—or, for that matter, photos of themselves in various states of undress? Want to see what your new friend used to look like in his drugged-out days? Want to find mobsters in the Witness Protection program, or spies in deep cover? The possibilities are nearly limitless.

To be sure, doing face recognition right takes an immense amount of computing power. The tool in Picasa is slow—on my laptop, it crunches for minutes. So for the time being, it may be too expensive to do it well for the whole Web. But face recognition has Moore's law, one of the most powerful laws in computing, on its side: Every year, as processor speed per dollar doubles, it'll get twice as cheap to do. Sooner or later, mass face recognition—perhaps even in real time, which would allow for recognition on security and video feeds—will roll out.

Facial recognition is especially significant because it'll create a kind of privacy discontinuity. We're used to a public semi-anonymity—while we know we may be spotted in a club or on the street, it's unlikely that we will be. But as security-camera and camera-phone pictures become searchable by face, that expectation will slip away. Shops with cameras facing the doors—and aisles—will be able to watch precisely where individual customers wander, what they pick up, and how this correlates with the data already collected about them by firms like Acxiom. And this powerful set of data—where you go and what you do, as indicated by where your face shows up in the bitstream—can be used to provide ever more custom-tailored experiences.

It's not just people that will be easier than ever to track. It's also individual objects—what some researchers are calling the "Internet of things."

As sci-fi author William Gibson once said, "The future is already here—it's just not very evenly distributed." It shows up in some places before others. And one of the places this particular aspect of the future has shown up first, oddly enough, is the Coca-Cola Village Amusement Park, a holiday village, theme park, and marketing event that opens seasonally in Israel. Sponsored by Facebook and Coke, the teenagers attending the park in the summer of 2010 were given bracelets containing a tiny piece of circuitry that allowed them to Like real-world objects. Wave the bracelet at the entrance to a ride, for example, and a status update posted to your account testifies that you're about to embark. Take a picture of your friends with a special camera and wave the bracelet at it, and the photo's posted with your identity already tagged.

Embedded in each bracelet is a radio-frequency identification (RFID) chip. RFID chips don't need batteries, and there's only one way to use them: call-and-response. Provide a little wireless electromagnetic power, and the chip chirps out a unique identifying code. Correlate the code with, say, a Facebook account, and you're in business. A single chip can cost as little as $.07, and they'll cost far less in the years to come.

Suddenly it's possible for businesses to track each individual object they make across the globe. Affix a chip to an individual car part, and you can watch as the part travels to the car factory, gets assembled into a car, and makes its way to the show floor and then someone's garage. No more inventory shrinkage, no more having to recall whole models of products because of the errors of one factory.

Conversely, RFID provides a framework by which a home could automatically inventory every object inside it—and track which objects are in which rooms. With a powerful enough signal, RFID could be a permanent solution to the lost-keys problem—and bring us face-to-face with what *Forbes* writer Reihan Salam calls "the powerful promise of a real world that can be indexed and organized as cleanly and coherently as Google has indexed and organized the Web."

This phenomenon is called ambient intelligence. It's based on a simple observation: The items you own, where you put them, and what you do with them is, after all, a great signal about what kind of person you are and what kind of preferences you have. "In the near future," writes a team of ambient intelligence experts led by David Wright, "every manufactured product—our clothes, money, appliances, the paint on our walls, the carpets on our floors, our cars, everything—will be embedded with intelligence, networks of tiny sensors and actuators, which some have termed 'smart dust.'"

And there's a third set of powerful signals that is getting cheaper and cheaper. In 1990, it cost about $10 to sequence a single base pair—one "letter"—of DNA. By 1999, that number had dropped to $.90. In 2004, it crossed the $.01 threshold, and now, as I write in 2010, it costs one ten-thousandth of $.01. By the time this book comes out, it'll undoubtedly cost exponentially less. By some point mid-decade, we ought to be able to sequence any random whole human genome for less than the cost of a sandwich.

It seems like something out of *Gattaca*, but the allure of adding this data to our profiles will be strong. While it's increasingly clear that our DNA doesn't determine everything about us—other cellular information sets, hormones, and our environment play a large role—there are undoubtedly numerous correlations between genetic material and behavior to be made. It's not just that we'll be able to predict and avert upcoming health issues with far greater accuracy—though that alone will be enough to get many of us in the door. By adding together DNA and behavioral data—like the location information from iPhones or the text of Facebook status updates—an enterprising scientist could run statistical regression analysis on an entire society.

In all this data lie patterns yet undreamed of. Properly harnessed, it will fuel a level of filtering acuity that's hard to imagine—a world in which nearly all of our objective experience is quantified, captured, and used to inform our environments. The biggest challenge, in fact, may be thinking of the right questions to ask of these enormous flows of binary digits. And increasingly, code will learn to ask these questions itself.

The End of Theory

In December 2010, researchers at Harvard, Google, *Encyclopædia Britannica*, and the *American Heritage Dictionary* announced the results of a four-year joint effort. The team had built a database spanning the entire contents of over five hundred years' worth of books—5.2 million books in total, in English, French,

Chinese, German, and other languages. Now any visitor to Google's "N-Gram viewer" page can query it and watch how phrases rise and fall in popularity over time, from neologism to the long fade into obscurity. For the researchers, the tool suggested even grander possibilities—a "quantitative approach to the humanities," in which cultural changes can be scientifically mapped and measured.

The initial findings suggest how powerful the tool can be. By looking at the references to previous dates, the team found that "humanity is forgetting its past faster with each passing year." And, they argued, the tool could provide "a powerful tool for automatically identifying censorship and propaganda" by identifying countries and languages in which there was a statistically abnormal absence of certain ideas or phrases. Leon Trotsky, for example, shows up far less in midcentury Russian books than in English or French books from the same time.

The project is undoubtedly a great service to researchers and the casually curious public. But serving academia probably wasn't Google's only motive. Remember Larry Page's declaration that he wanted to create a machine "that can understand anything," which some people might call artificial intelligence? In Google's approach to creating intelligence, the key is data, and the 5 million digitized books contain an awful lot of it. To grow your artificial intelligence, you need to keep it well fed.

To get a sense of how this works, consider Google Translate, which can now do a passable job translating automatically among nearly sixty languages. You might imagine that Translate was built with a really big, really sophisticated set of translating dictionaries, but you'd be wrong. Instead, Google's engineers took a probabilistic approach: They built software that could identify which words tended to appear in connection with which, and then sought out large chunks of data that were available in multiple languages to train the software on. One of the largest chunks was patent and trademark filings, which are useful because they all say the same thing, they're in the public domain, and they have to be filed globally in scores of different languages. Set loose on a hundred thousand patent applications in English and French, Translate could determine that when *word* showed up in the English document, *mot* was likely to show up in the corresponding French paper. And as users correct Translate's work over time, it gets better and better.

What Translate is doing with foreign languages Google aims to do with just about everything. Cofounder Sergey Brin has expressed his interest in plumbing genetic data. Google Voice captures millions of minutes of human speech, which engineers are hoping they can use to build the next generation of speech-recognition software. Google Research has captured most of the scholarly articles in the world. And of course, Google's search users pour billions of queries into the machine every day, which provide another rich vein of cultural information. If you had a secret plan to vacuum up an entire civilization's data and use it to build artificial intelligence, you couldn't do a whole lot better.

As Google's protobrain increases in sophistication, it'll open up remarkable new possibilities. Researchers in Indonesia can benefit from the latest papers in Stanford (and vice versa) without waiting for translation delays. In a matter of a few years, it may be possible to have an automatically translated voice conversation with someone speaking a different language, opening up whole new channels of cross-cultural communication and understanding.

But as these systems become increasingly "intelligent," they also become harder to control and understand. It's not quite right to say they take on a life of their own—ultimately, they're still just code. But they reach a level of complexity at which even their programmers can't fully explain any given output.

This is already true to a degree with Google's search algorithm. Even to its engineers, the workings of the algorithm are somewhat mysterious. "If they opened up the mechanics," says search expert Danny Sullivan, "you still wouldn't understand it. Google could tell you all two hundred signals it uses and what the code is and you wouldn't know what to do with them." The core software engine of Google search is hundreds of thousands of lines of code. According to one Google employee I talked to who had spoken to the search team, "The team tweaks and tunes, they don't really know what works or why it works, they just look at the result."

Google promises that it doesn't tilt the deck in favor of its own products. But the more complex and "intelligent" the system gets, the harder it'll be to tell. Pinpointing where bias or error exists in a human brain is difficult or impossible—there are just too many neurons and connections to narrow it down to a single malfunctioning chunk of tissue. And as we rely on intelligent systems like Google's more, their opacity could cause real problems—like the still-mysterious machine-driven "flash crash" that caused the Dow to drop 600 points in a few minutes on May 6, 2010.

In a provocative article in *Wired*, editor-in-chief Chris

Anderson argued that huge databases render scientific theory itself obsolete. Why spend time formulating human-language hypotheses, after all, when you can quickly analyze trillions of bits of data and find the clusters and correlations? He quotes Peter Norvig, Google's research director: "All models are wrong, and increasingly you can succeed without them." There's plenty to be said for this approach, but it's worth remembering the downside: Machines may be able to see results without models, but humans can't understand without them. There's value in making the processes that run our lives comprehensible to the humans who, at least in theory, are their beneficiaries.

Supercomputer inventor Danny Hillis once said that the greatest achievement of human technology is tools that allow us to create more than we understand. That's true, but the same trait is also the source of our greatest disasters. The more the code driving personalization comes to resemble the complexity of human cognition, the harder it'll be to understand why or how it's making the decisions it makes. A simple coded rule that bars people from one group or class from certain kinds of access is easy to spot, but when the same action is the result of a swirling mass of correlations in a global supercomputer, it's a trickier problem. And the result is that it's harder to hold these systems and their tenders accountable for their actions.

No Such Thing as a Free Virtual Lunch

In January 2009, if you were listening to one of twenty-five radio stations in Mexico, you might have heard the accordion ballad "El más grande enemigo." Though the tune is polka-ish and cheery, the lyrics depict a tragedy: a migrant seeks to illegally cross the border, is betrayed by his handler, and is left in the blistering desert sun to die. Another song from the *Migra corridos* album tells a different piece of the same sad tale:

> To cross the border
> I got in the back of a trailer
> There I shared my sorrows
> With forty other immigrants
> I was never told
> That this was a trip to hell.

If the lyrics aren't exactly subtle about the dangers of crossing the border, that's the point. *Migra corridos* was produced by a contractor working for the U.S. Border Control, as part of a campaign to stem the tide of immigrants along the border. The song is a prime example of a growing trend in what marketers delicately call "advertiser-funded media," or AFM.

Product placement has been in vogue for decades, and AFM is its natural next step. Advertisers love product placement because in a media environment in which it's harder and harder to get people to pay attention to anything—especially ads—it provides a kind of loophole. You can't fast-forward past product placement. You can't miss it without missing some of the actual content. AFM is just a natural extension of the same logic: Media have always been vehicles for selling products, the argument goes, so why not just cut out the middleman and have product makers produce the content themselves?

In 2010, Walmart and Procter & Gamble announced a partnership to produce *Secrets of the Mountain* and *The Jensen Project*, family movies that will feature characters using the companies' products throughout. Michael Bay, the director of *Transformers*, has started a new company called the Institute, whose tagline is "Where Brand Science Meets Great Storytelling." *Hansel and Gretel in 3-D*, its first feature production, will be specially crafted to provide product-placement hooks throughout.

Now that the video-game industry is far more profitable than the movie industry, it provides a huge opportunity for in-game advertising and product placement as well. Massive Incorporated, a game advertising platform acquired by Microsoft for $200 million to $400 million, has placed ads on in-game billboards and city walls for companies like Cingular and McDonald's, and has the capacity to track which individual users saw which advertisements for how long. Splinter Cell, a game by UBIsoft, works placement for products like Axe deodorant into the architecture of the cityscape that characters travel through.

Even books aren't immune. *Cathy's Book*, a young-adult title published in September 2006, has its heroine applying "a killer coat of Lipslicks in 'Daring.'" That's not a coincidence—*Cathy's Book* was published by Procter & Gamble, the corporate owner of Lipslicks.

If the product placement and advertiser-funded media industries continue to grow, personalization will offer whole new vistas of possibility. Why name-drop Lipslicks when your reader is more likely to buy Cover Girl? Why have a video-game chase scene through Macy's when the guy holding the controller is more of an Old Navy type? When software engineers talk about architecture, they're usually talking metaphorically. But as people spend more of their time in virtual, personalizable places, there's no reason that these worlds can't change to suit users' preferences. Or, for that matter, a corporate sponsor's.

A Shifting World

The enriched psychological models and new data flows measuring everything from heart rate to music choices open up new frontiers for online personalization, in which what changes isn't just a choice of products or news clips, but the look and feel of the site on which they're displayed.

Why *should* Web sites look the same to every viewer or customer? Different people don't respond only to different products—they respond to different design sensibilities, different colors, even different types of product descriptions. It's easy enough to imagine a Walmart Web site with softened edges and warm pastels for some customers and a hard-edged, minimalist design for others. And once that capacity exists, why stick with just one design per customer? Maybe it's best to show me one side of the Walmart brand when I'm angry and another when I'm happy.

This kind of approach isn't a futuristic fantasy. A team led by John Hauser at MIT's business school has developed the basic techniques for what they call Web site morphing, in which a shopping site analyzes users' clicks to figure out what kinds of information and styles of presentation are most effective and then adjusts the layout to suit a particular user's cognitive style. Hauser estimates that Web sites that morph can increase "purchase intentions" by 21 percent. Industrywide, that's worth billions. And what starts with the sale of consumer products won't end there: News and entertainment sources that morph ought to enjoy an advantage as well.

On one hand, morphing makes us feel more at home on the Web. Drawing from the data we provide, every Web site can feel like an old friend. But it also opens the door to a strange, dreamlike world, in which our environment is constantly rearranging itself behind our backs. And like a dream, it may be less and less possible to share with people outside of it—that is, everyone else.

Thanks to augmented reality, that experience may soon be par for the course offline as well.

"On the modern battlefield," Raytheon Avionics manager Todd Lovell told a reporter, "there is way more data out there than most people can use. If you are just trying to see it all through your eyes and read it in bits and bites, you're never going to understand it. So the key to the modern technology is to take all that data and turn it into useful information that the pilot can recognize very quickly and act upon." What Google does for online information, Lovell's Scorpion project aims to do for the real world.

Fitting like a monocle over one of a jet pilot's eyes, the Scorpion display device annotates what a pilot sees in real time. It color-codes potential threats, highlights when and where the aircraft has a missile lock, assists with night vision, and reduces the need for pilots to look at a dashboard in an environment where every microsecond matters. "It turns the whole world into a display," jet pilot Paul Mancini told the Associated Press.

This is augmented-reality technology, and it's moving rapidly from the cockpits of jet planes to consumer devices that can tune out the noise and turn up the signal of everyday life. Using your iPhone camera and an app developed by Yelp, the restaurant recommendation service, you can see eateries' ratings haphazardly displayed over their real-world storefronts. A new kind of noise-canceling headphone can sense and amplify human voices while tuning other street or airplane noise down to a whisper. The Meadowlands football stadium is spending $100 million on new applications that give fans who attend games in person the ability to slice and dice the game in real time, view key statistics as they happen, and watch the action unfold from a variety of angles—the full high-information TV experience overlaid on a real game.

At DARPA, the defense research and development agency, technologies are being developed that make Scorpion look positively quaint. Since 2002, DARPA has been pushing forward research in what it calls augmented cognition, or AugCog, which uses cognitive neuroscience and brain imaging to figure out how best to route important information into the brain. AugCog begins with the premise that there are basic limits as to how many tasks a person can juggle at a time, and that "this capacity itself may fluctuate from moment to moment depending on a host of factors including mental fatigue, novelty, boredom and stress."

By monitoring activity in brain areas associated with memory, decision making, and the like, AugCog devices can figure out how to make sure to highlight the information that most matters. If you're absorbing as much visual input as you can, the system might decide to send an audio alert instead. One trial, according to the *Economist*, gave users of an AugCog device a 100 percent improvement in recall and a 500 percent increase in working memory. And if it sounds far-fetched, just remember: The folks at DARPA also helped invent the Internet.

Augmented reality is a booming field, and Gary Hayes, a personalization and augmented-reality expert in Australia, sees at least sixteen different ways it could be used to provide services and make money. In his vision, guide companies could offer augmented reality tours, in which information about buildings, museum artifacts, and streets is superimposed on the environs. Shoppers could use phone apps to immediately get readouts on products they're interested in—including what the objects cost elsewhere. (Amazon.com already provides a rudimentary version of this service.) Augmented reality games could layer clues into real-world environments.

Augmented-reality tech provides value, but it also provides an opportunity to reach people with new attention-getting forms of advertising. For a price, digital sportscasts are already capable of layering corporate logos onto football fields. But this new technology offers the opportunity to do that in a personalized way in the real world: You turn on the app to, say, help find a friend in a crowd, and projected onto a nearby building is a giant Coke ad featuring your face and your name.

And when you combine the personalized filtering of what we see and hear with, say, face recognition, things get pretty interesting: You begin to be able to filter not just information, but people.

As the cofounder of OkCupid, one of the Web's most popular dating sites, Chris Coyne has been thinking about filtering for people for a while. Coyne speaks in an energetic, sincere manner, furrowing his brows when he's thinking and waving his hands to illustrate. As a math major, he got interested in how to use algorithms to solve problems for people.

"There are lots of ways you can use math to do things that turn a profit," he told me over a steaming bowl of bibimbap in New York's Koreatown. Many of his classmates went off to high-paid jobs at hedge funds. "But," he said, "what we were interested in was using it to make people happy." And what better way to make people happy than to help them fall in love?

The more Coyne and his college hallmates Sam Yeager and Max Krohn looked at other dating sites, the more annoyed they got: It was clear that other dating sites were more interested in getting people to pay for credits than to hook up. And once you did pay, you'd often see profiles of people who were no longer on the site or who would never write you back.

Coyne and his team decided to approach the problem with math. The service would be free. Instead of offering a one-size-fits-all solution, they'd use number crunching to develop a personalized matching algorithm for each person on the site. And just as Google optimizes for clicks, they'd do everything they could to maximize the likelihood of real conversations—if you could solve for that, they figured, profits would follow. In essence, they built a modern search engine for mates.

When you log on to OkCupid, you're asked a series of questions about yourself. Do you believe in God? Would you ever participate in a threesome? Does smoking disgust you? Would you sleep with someone on the first date? Do you have an STD? (Answer yes, and you get sent to another site.) You also indicate how you'd like a prospective partner to answer the same questions and how important their answers are to you. Using these questions, OkCupid builds a custom-weighted equation to figure out your perfect match. And when you search for people in your area, it uses the same algorithm to rank the likelihood of your getting along. OkCupid's powerful cluster of servers can rank ten thousand people with a two-hundred-question match model and return results in less than a tenth of a second.

They have to, because OkCupid's traffic is booming. Hundreds of thousands of answers to poll questions flow into their system each night. Thousands of new users sign up each day. And the system is getting better and better.

Looking into the future, Coyne told me, you'll have people walking around with augmented displays. He described a guy on a night out: You walk into a bar, and a camera immediately scans the faces in the room and matches them against OkCupid's databases. "Your accessories can say, that girl over there is an eighty-eight percent match. That's a dream come true!"

Vladimir Nabokov once commented that "reality" is "one of the few words that mean nothing without quotes." Coyne's vision may soon be our "reality." There's tremendous promise in this vision: Surgeons who never miss a suture, soldiers who never imperil civilians, and everywhere a more informed, information-dense world. But there's also danger: Augmented reality represents the end of naive empiricism, of the world as we see it, and the beginning of something far more mutable and weird: a real-world filter bubble that will be increasingly difficult to escape.

Losing Control

There's plenty to love about this ubiquitously personalized future.

Smart devices, from vacuum cleaners to lightbulbs to picture frames, offer the promise that our environments will be exactly the way we want them, wherever we are. In the near future, ambient-intelligence expert David Wright suggests, we might even carry our room-lighting preferences with us; when there are multiple people in a room, a consensus could be automatically reached by averaging preferences and weighting for who's the host.

AugCog-enabled devices will help us track the data streams that we consider most important. In some situations—say, medical or fire alerts that find ways to escalate until they capture our attention—they could save lives. And while brain-wave-reading AugCog is probably some way off for the masses, consumer variants of the basic concept are already being put into place. Google's Gmail Priority Inbox, which screens e-mails and highlights the ones it assesses as more important, is an early riff on the theme. Meanwhile, augmented-reality filters offer the possibility of an annotated and hyperlinked reality, in which what we see is infused with information that allows us to work better, assimilate information more quickly, and make better decisions.

That's the good side. But there's always a bargain in personalization: In exchange for convenience, you hand over some privacy and control to the machine.

As personal data become more and more valuable, the behavioral data market described in chapter 1 is likely to explode. When a clothing company determines that knowing your favorite color produces a $5 increase in sales, it has an economic basis for pricing that data point—and for other Web sites to find reasons to ask you. (While OkCupid is mum about its business model, it likely rests on offering advertisers the ability to target its users based on the hundreds of personal questions they answer.)

While many of these data acquisitions will be legitimate, some won't be. Data are uniquely suited to gray-market activities, because they need not carry any trace of where they have come from or where they have been along the way. Wright calls this data laundering, and it's already well under way: Spyware and spam companies sell questionably derived data to middlemen, who then add it to the databases powering the marketing campaigns of major corporations.

Moreover, because the transformations applied to your data are often opaque, it's not always clear exactly what decisions are being made on your behalf, by whom, or to what end. This matters plenty when we're talking about information streams, but it matters even more when this power is infused into our sensory apparatus itself.

In 2000, Bill Joy, the Sun Microsystems cofounder, wrote a piece for *Wired* magazine titled "Why the Future Doesn't Need Us." "As society and the problems that face it become more and more complex and machines become more and more intelligent," he wrote, "people will let machines make more of their decisions for them, simply because machine-made decisions will bring better results than man-made ones."

That may often be the case: Machine-driven systems do provide significant value. The whole promise of these technologies is that they give us more freedom and more control over our world—lights that respond to our whims and moods, screens and overlays that allow us to attend only to the people we want to, so that we don't have to do the busywork of living. The irony is that they offer this freedom and control by taking it away. It's one thing when a remote control's array of buttons elides our ability to do something basic like flip the channels. It's another thing when what the remote controls is our lives.

It's fair to guess that the technology of the future will work about as well as the technology of the past—which is to say, well enough, but not perfectly. There will be bugs. There will be dislocations and annoyances. There will be breakdowns that cause us to question whether the whole system was worth it in the first place. And we'll live with the threat that systems made to support us will be turned against us—that a clever hacker who cracks the baby monitor now has a surveillance device, that someone who can interfere with what we see can expose us to danger. The more power we have over our own environments, the more power someone who assumes the controls has over us.

That is why it's worth keeping the basic logic of these systems in mind: You don't get to create your world on your own. You live in an equilibrium between your own desires and what the market will bear. And while in many cases this provides for healthier, happier lives, it also provides for the commercialization of everything—even of our sensory apparatus itself. There are few things uglier to contemplate than AugCog-enabled ads that escalate until they seize control of your attention.

We're compelled to return to Jaron Lanier's question: For whom do these technologies work? If history is any guide, we may not be the primary customer. And as technology gets better and better at directing our attention, we need to watch closely what it is directing our attention toward.

8

Escape from the City of Ghettos

> In order to find his own self, [a person] also needs
> to live in a milieu where the possibility of many
> different value systems is explicitly recognized and
> honored. More specifically, he needs a great variety
> of choices so that he is not misled about the nature
> of his own person.
>
> —*Christopher Alexander* et al., *A Pattern Language*

In theory, there's never been a structure more capable of allowing all of us to shoulder the responsibility for understanding and managing our world than the Internet. But in practice, the Internet is headed in a different direction. Sir Tim Berners-Lee, the creator of the World Wide Web, captured the gravity of this threat in a recent call to arms in the pages of *Scientific American* titled "Long Live the Web." "The Web as we know it," he wrote, "is being threatened. . . . Some of its most successful inhabitants have begun to chip away at its principles. Large social-networking sites are walling off information posted by their users from the rest of the Web. . . . Governments—totalitarian and democratic alike—are monitoring people's on-line habits, endangering important human rights. If we, the

Web's users, allow these and other trends to proceed unchecked, the Web could be broken into fragmented islands."

In this book, I've argued that the rise of pervasive, embedded filtering is changing the way we experience the Internet and ultimately the world. At the center of this transformation is the fact that for the first time it's possible for a medium to figure out who you are, what you like, and what you want. Even if the personalizing code isn't always spot-on, it's accurate enough to be profitable, not just by delivering better ads but also by adjusting the substance of what we read, see, and hear.

As a result, while the Internet offers access to a dazzling array of sources and options, in the filter bubble we'll miss many of them. While the Internet can give us new opportunities to grow and experiment with our identities, the economics of personalization push toward a static conception of personhood. While the Internet has the potential to decentralize knowledge and control, in practice it's concentrating control over what we see and what opportunities we're offered in the hands of fewer people than ever before.

Of course, there are some advantages to the rise of the personalized Internet. I enjoy using Pandora, Netflix, and Facebook as much as the next person. I appreciate Google's shortcuts through the information jungle (and couldn't have written this book without them). But what's troubling about this shift toward personalization is that it's largely invisible to users and, as a result, out of our control. We are not even aware that we're seeing increasingly divergent images of the Internet. The Internet may know who we are, but we don't know who it thinks we are or how it's using that information. Technology designed to give us more control over our lives is actually taking control away.

Ultimately, Sun Microsystems cofounder Bill Joy told me, information systems have to be judged on their public outcomes. "If what the Internet does is spread around a lot of information, fine, but what did that cause to happen?" he asked. If it's not helping us solve the really big problems, what good is it? "We really need to address the core issues: climate change, political instability in Asia and the Middle East, demographic problems, and the decline of the middle class. In the context of problems of this magnitude, you'd hope that a new constituency would emerge, but there's a distraction overlay—false issues, entertainment, gaming. If our system, with all the freedom of choice, is not addressing the problems, something's wrong."

Something *is* wrong with our media. But the Internet isn't doomed, for a simple reason: This new medium is nothing if not plastic. Its great strength, in fact, is its capacity for change. Through a combination of individual action, corporate responsibility, and governmental regulation, it's still possible to shift course.

"We create the Web," Sir Tim Berners-Lee wrote. "We choose what properties we want it to have and not have. It is by no means finished (and it's certainly not dead)." It's still possible to build information systems that introduce us to new ideas, that push us in new ways. It's still possible to create media that show us what we don't know, rather than reflecting what we do. It's still possible to erect systems that don't trap us in an endless loop of self-flattery about our own interests or shield us from fields of inquiry that aren't our own.

First, however, we need a vision—a sense of what to aim for.

The Mosaic of Subcultures

In 1975, architect Christopher Alexander and a team of colleagues began publishing a series of books that would change the face of urban planning, design, and programming. The most famous volume, *A Pattern Language*, is a guidebook that reads like a religious text. It's filled with quotes and aphorisms and hand-drawn sketches, a bible guiding devotees toward a new way of thinking about the world.

The question that had consumed Alexander and his team during eight years of research was the question of why some places thrived and "worked" while others didn't—why some cities and neighborhoods and houses flourished, while others were grim and desolate. The key, Alexander argued, was that design has to fit its literal and cultural context. And the best way to ensure that, they concluded, was to use a "pattern language," a set of design specifications for human spaces.

Even for nonarchitects, the book is an entrancing read. There's a pattern that describes the ideal nook for kids (the ceiling should be between 2 feet 6 inches and 4 feet high), and another for High Places "where you can look down and survey your world." "Every society which is alive and whole," Alexander wrote, "will have its own unique and distinct pattern language."

Some of the book's most intriguing sections illuminate the patterns that successful cities are built on. Alexander imagines two metropolises—the "heterogeneous city," where people are mixed together irrespective of lifestyle and background, and the "city of ghettos," where people are grouped together tightly by category. The heterogeneous city "seems rich," Alexander writes, but "actually it dampens all significant variety, and arrests most of the possibilities for differentiation." Though there's a diverse mix of peoples and cultures, all of the parts of the city are diverse in the same way. Shaped by the lowest common cultural denominators, the city looks the same everywhere you go.

Meanwhile, in the city of ghettos, some people get trapped in the small world of a single subculture that doesn't really represent who they are. Without connections and overlap between communities, subcultures that make up the city don't evolve. As a result, the ghettos breed stagnation and intolerance.

But Alexander offers a third possibility: a happy medium between closed ghettos and the undifferentiated mass of the heterogeneous city. He called it the mosaic of subcultures. In order to achieve this kind of city, Alexander explains, designers should encourage neighborhoods with cultural character, "but though these subcultures must be sharp and distinct and separate, they must not be closed; they must be readily accessible to one another, so that a person can move easily from one to another, and can settle in the one which suits him best." Alexander's mosaic is based on two premises about human life: First, a person can only fully become him- or herself in a place where he or she "receives support for his idiosyncrasies from the people and values which surround him." And second, as the quotation at the beginning of this chapter suggests, you have to see lots of ways of living in order to choose the best life for yourself. This is what the best cities do: They cultivate a vibrant array of cultures and allow their citizens to find their way to the neighborhoods and traditions in which they're most at home.

Alexander was writing about cities, but what's beautiful about *A Pattern Language* is that it can be applied to any space in which humans gather and live—including the Internet. Online communities and niches are important. They're the places where new ideas and styles and themes and even languages get formed and tested. They're the places where we can feel most at home. An Internet built like the heterogeneous city described by Alexander wouldn't be a very pleasant place to be—a whirling chaos of facts and ideas and communications. But by the same token, nobody wants to live in the city of ghettos—and that's where personalization, if it's too acute, will take us. At its worst, the filter bubble confines us to our own information neighborhood, unable to see or explore the rest of the enormous world of possibilities that exist online. We need our online urban planners to strike a balance between relevance and serendipity, between the comfort of seeing friends and the exhilaration of meeting strangers, between cozy niches and wide open spaces.

What Individuals Can Do

Social-media researcher danah boyd was right when she warned that we are at risk of the "psychological equivalent of obesity." And while creating a healthy information diet requires action on the part of the companies that supply the food, that doesn't work unless we also change our own habits. Corn syrup vendors aren't likely to change their practices until consumers demonstrate that they're looking for something else.

Here's one place to start: Stop being a mouse.

On an episode of the radio program *This American Life,* host Ira Glass investigates how to build a better mousetrap. He talks to Andy Woolworth, the man at the world's largest mousetrap manufacturer who fields ideas for new trap designs. The proposed ideas vary from the impractical (a trap that submerges the mouse in antifreeze, which then needs to be thrown out by the bucket) to the creepy (a design that kills rodents using, yes, gas pellets).

But the punch line is that they're all unnecessary. Woolworth has an easy job, because the existing traps are very cheap and work within a day 88 percent of the time. Mousetraps work because mice generally establish a food-seeking route within ten feet of where they are, returning to it up to thirty times a day. Place a trap in its vicinity, and chances are very good that you'll catch your mouse.

Most of us are pretty mouselike in our information habits. I admittedly am: There are three or four Web sites that I check frequently each day, and I rarely vary them or add new ones to my repertoire. "Whether we live in Calcutta or San Francisco," Matt Cohler told me, "we all kinda do the same thing over and over again most of the time. And jumping out of that recursion loop is not easy to do." Habits are hard to break. But just as you notice more about the place you live when you take a new route to work, varying your path online dramatically increases your likelihood of encountering new ideas and people.

Just by stretching your interests in new directions, you give the personalizing code more breadth to work with. Someone who shows interest in opera and comic books and South

African politics and Tom Cruise is harder to pigeonhole than someone who just shows interest in one of those things. And by constantly moving the flashlight of your attention to the perimeter of your understanding, you enlarge your sense of the world.

Going off the beaten track is scary at first, but the experiences we have when we come across new ideas, people, and cultures are powerful. They make us feel human. Serendipity is a shortcut to joy.

For some of the "identity cascade" problems discussed in chapter 5, regularly erasing the cookies your Internet browser uses to identify who you are is a partial cure. Most browsers these days make erasing cookies pretty simple—you just select Options or Preferences and then choose Erase cookies. And many personalized ad networks are offering consumers the option to opt out. I'm posting an updated and more detailed list of places to opt out on the Web site for this book, www .thefilterbubble.com.

But because personalization is more or less unavoidable, opting out entirely isn't a particularly viable route for most of us. You can run all of your online activities in an "incognito" window, where less of your personal information is stored, but it'll be increasingly impractical—many services simply won't work the way they're supposed to. (This is why, as I describe below, I don't think the Do Not Track list currently under consideration by the FTC is a viable strategy.) And of course, Google personalizes based on your Internet address, location, and a number of other factors even if you're entirely logged out and on a brand-new laptop.

A better approach is to choose to use sites that give users

more control and visibility over how their filters work and how they use your personal information.

For example, consider the difference between Twitter and Facebook. In many ways, the two sites are very similar. They both offer people the opportunity to share blips of information and links to videos, news, and photographs. They both offer the opportunity to hear from the people you want to hear from and screen out the people you don't.

But Twitter's universe is based on a few very simple, mostly transparent rules—what one Twitter supporter called "a thin layer of regulation." Unless you go out of your way to lock your account, everything you do is public to everyone. You can subscribe to anyone's feed that you like without their permission, and then you see a time-ordered stream of updates that includes everything everyone you're following says.

In comparison, the rules that govern Facebook's information universe are maddeningly opaque and seem to change almost daily. If you post a status update, your friends may or may not see it, and you may or may not see theirs. (This is true even in the Most Recent view that many users assume shows all of the updates—it doesn't.) Different types of content are likely to show up at different rates—if you post a video, for example, it's more likely to be seen by your friends than a status update. And the information you share with the site itself is private one day and public the next. There's no excuse, for example, for asking users to declare which Web sites they're "fans" of with the promise that it'll be shown only to their friends, and then releasing that information to the world, as Facebook did in 2009.

Because Twitter operates on the basis of a few simple, easily understandable rules, it's also less susceptible to what venture capitalist Brad Burnham (whose Union Square Ventures was Twitter's primary early investor) calls the tyranny of the default. There's great power in setting the default option when people are given a choice. Dan Ariely, the behavioral economist, illustrates the principle with a chart showing organ donation rates in different European countries. In England, the Netherlands, and Austria, the rates hover around 10 percent to 15 percent, but in France, and Belgium, donation rates are in the high 90s. Why? In the first set of countries, you have to check a box giving permission for your organs to be donated. In the second, you have to check a box to say you *won't* give permission.

If people will let defaults determine the fate of our friends who need lungs and hearts, we'll certainly let them determine how we share information a lot of the time. That's not because we're stupid. It's because we're busy, have limited attention with which to make decisions, and generally trust that if everyone else is doing something, it's OK for us to do it too. But this trust is often misplaced. Facebook has wielded this power with great intentionality—shifting the defaults on privacy settings in order to encourage masses of people to make their posts more public. And because software architects clearly understand the power of the default and use it to make their services more profitable, their claim that users *can* opt out of giving their personal information seems somewhat disingenuous. With fewer rules and a more transparent system, there are fewer defaults to set.

Facebook's PR department didn't return my e-mails requesting an interview (perhaps because MoveOn's critical view of Facebook's privacy practices is well known). But it would probably argue that it gives its users far more choice and control about how they use the service than Twitter does. And it's true that Facebook's options control panel lists scores of different options for Facebook users.

But to give people control, you have to make clearly evident what the options are, because options largely exist only to the degree that they're perceived. This is the problem many of us used to face in programming our VCRs: The devices had all sorts of functions, but figuring out how to make them do anything was an afternoon-long exercise in frustration. When it comes to important tasks like protecting your privacy and adjusting your filters online, saying that you can figure it out if you read the manual for long enough isn't a sufficient answer.

In short, at the time of this writing, Twitter makes it pretty straightforward to manage your filter and understand what's showing up and why, whereas Facebook makes it nearly impossible. All other things being equal, if you're concerned about having control over your filter bubble, better to use services like Twitter than services like Facebook.

We live in an increasingly algorithmic society, where our public functions, from police databases to energy grids to schools, run on code. We need to recognize that societal values about justice, freedom, and opportunity are embedded in how code is written and what it solves for. Once we understand that, we can begin to figure out which variables we care about and imagine how we might solve for something different.

For example, advocates looking to solve the problem of political gerrymandering—the backroom process of carving up electoral districts to favor one party or another—have long suggested that we replace the politicians involved with software. It sounds pretty good: Start with some basic principles, input population data, and out pops a new political map. But it doesn't necessarily solve the basic problem, because what the algorithm solves for has political consequences: Whether the software aims to group by cities or ethnic groups or natural boundaries can determine which party keeps its seats in Congress and which doesn't. And if the public doesn't pay close attention to what the algorithm is doing, it could have the opposite of the intended effect—sanctioning a partisan deal with the imprimatur of "neutral" code.

In other words, it's becoming more important to develop a basic level of algorithmic literacy. Increasingly, citizens will have to pass judgment on programmed systems that affect our public and national life. And even if you're not fluent enough to read through thousands of lines of code, the building-block concepts—how to wrangle variables, loops, and memory—can illuminate how these systems work and where they might make errors.

Especially at the beginning, learning the basics of programming is even more rewarding than learning a foreign language. With a few hours and a basic platform, you can have that "Hello, World!" experience and start to see your ideas come alive. And within a few weeks, you can be sharing these ideas with the whole Web. Mastery, as in any profession, takes much longer, but the payoff for a limited investment in coding is fairly large: It doesn't take long to become literate enough to understand what most basic bits of code are doing.

Changing our own behavior is a part of the process of bursting the filter bubble. But it's of limited use unless the companies that are propelling personalization forward change as well.

What Companies Can Do

It's understandable that, given their meteoric rises, the Googles and Facebooks of the online world have been slow to realize their responsibilities. But it's critical that they recognize their public responsibility soon. It's no longer sufficient to say that the personalized Internet is just a function of relevance-seeking machines doing their job.

The new filterers can start by making their filtering systems more transparent to the public, so that it's possible to have a discussion about how they're exercising their responsibilities in the first place.

As Larry Lessig says, "A political response is possible only when regulation is transparent." And there's more than a little irony in the fact that companies whose public ideologies revolve around openness and transparency are so opaque themselves.

Facebook, Google, and their filtering brethren claim that to reveal anything about their algorithmic processes would be to give away business secrets. But that defense is less convincing than it sounds at first. Both companies' primary advantage lies in the extraordinary number of people who trust them and use their services (remember lock-in?). According to Danny Sullivan's Search Engine Land blog, Bing's search results are "highly competitive" with Google's, but it has a fraction of its more powerful rival's users. It's not a matter of math that keeps Google ahead, but the sheer number of people who use it every day. PageRank and the other major pieces of Google's search engine are "actually one of the world's worst kept secrets," says Google fellow Amit Singhal.

Google has also argued that it needs to keep its search algorithm under tight wraps because if it was known it'd be easier to game. But open systems are harder to game than closed ones, precisely because everyone shares an interest in closing loopholes. The open-source operating system Linux, for example, is actually more secure and harder to penetrate with a virus than closed ones like Microsoft's Windows or Apple's OS X.

Whether or not it makes the filterers' products more secure or efficient, keeping the code under tight wraps does do one thing: It shields these companies from accountability for the decisions they're making, because the decisions are difficult to see from the outside. But even if full transparency proves impossible, it's possible for these companies to shed more light on how they approach sorting and filtering problems.

For one thing, Google and Facebook and other new media giants could draw inspiration from the history of newspaper ombudsmen, which became a newsroom topic in the mid-1960s.

Philip Foisie, an executive at the *Washington Post* company, wrote one of the most memorable memos arguing for the practice. "It is not enough to say," he suggested, "that our paper, as it appears each morning, is its own credo, that ultimately we are our own ombudsman. It has not proven to be, possibly cannot be. Even if it were, it would not be viewed as such. It is too much to ask the reader to believe that we are capable of being honest and objective about ourselves." The *Post* found his argument compelling, and hired its first ombudsman in 1970.

"We know the media is a great dichotomy," said the longtime *Sacramento Bee* ombudsman Arthur Nauman in a speech in 1994. On the one hand, he said, media has to operate as a successful business that provides a return on investment. "But on the other hand, it is a public trust, a kind of public utility. It is an institution invested with enormous power in the community, the power to affect thoughts and actions by the way it covers the news—the power to hurt or help the common good." It is this spirit that the new media would do well to channel. Appointing an independent ombudsman and giving the world more insight into how the powerful filtering algorithms work would be an important first step.

Transparency doesn't mean only that the guts of a system are available for public view. As the Twitter versus Facebook dichotomy demonstrates, it also means that individual users intuitively understand how the system works. And that's a necessary precondition for people to control and use these tools—rather than having the tools control and use us.

To start with, we ought to be able to get a better sense of who these sites think we are. Google claims to make this possible with a "dashboard"—a single place to monitor and manage all of this data. In practice, its confusing and multitiered design makes it almost impossible for an average user to navigate and understand. Facebook, Amazon, and other companies don't allow users to download a complete compilation of their data in the United States, though privacy laws in Europe force them to. It's an entirely reasonable expectation that data that users provide to companies ought to be available to us, and that this expectation is one that, according to the University of California at Berkeley, most Americans share. We ought to be able to say, "You're wrong. Perhaps I used to be a surfer, or a fan of comics, or a Democrat, but I'm not any more."

Knowing what information the personalizers have on us isn't enough. They also need to do a much better job explaining how they use the data—what bits of information are personalized, to what degree, and on what basis. A visitor to a personalized news site could be given the option of seeing how many other visitors were seeing which articles—even perhaps a color-coded visual map of the areas of commonality and divergence. Of course, this requires admitting to the user that personalization is happening in the first place, and there are strong reasons in some cases for businesses not to do so. But they're mostly commercial reasons, not ethical ones.

The Interactive Advertising Bureau is already pushing in this direction. An industry trade group for the online advertising community, the IAB has concluded that unless personalized ads disclose to users how they're personalized, consumers will get angry and demand federal regulation. So it's encouraging its members to include a set of icons on every ad to indicate what personal data the ad draws on and how to change or opt out of this feature set. As content providers incorporate the personalization techniques pioneered by direct marketers and advertisers, they should consider incorporating these safeguards as well.

Even then, sunlight doesn't solve the problem unless it's coupled with a focus in these companies on optimizing for different variables: more serendipity, a more humanistic and nuanced sense of identity, and an active promotion of public issues and cultivation of citizenship.

As long as computers lack consciousness, empathy, and intelligence, much will be lost in the gap between our actual selves and the signals that can be rendered into personalized environments. And as I discussed in chapter 5, personalization algorithms can cause identity loops, in which what the code knows about you constructs your media environment, and your media environment helps to shape your future preferences. This is an avoidable problem, but it requires crafting an algorithm that prioritizes "falsifiability," that is, an algorithm that aims to *dis*-prove its idea of who you are. (If Amazon harbors a hunch that you're a crime novel reader, for example, it could actively present you with choices from other genres to fill out its sense of who you are.)

Companies that hold great curatorial power also need to do more to cultivate public space and citizenship. To be fair, they're already doing some of this: Visitors to Facebook on November 2, 2010, were greeted by a banner asking them to indicate if they'd voted. Those who had voted shared this news with their friends; because some people vote because of social pressure, it's quite possible that Facebook increased the number of voters. Likewise, Google has been doing strong work to make information about polling locations more open and easily available, and featured its tool on its home page on the same day. Whether or not this is profit-seeking behavior (a "find your polling place" feature would presumably be a terrific place for political advertising), both projects drew the attention of users toward political engagement and citizenship.

A number of the engineers and technology journalists I talked to raised their eyebrows when I asked them if personalizing algorithms could do a better job on this front. After all, one said, who's to say what's important? For Google engineers to place a value on some kinds of information over others, another suggested, would be unethical—though of course this is precisely what the engineers themselves do all the time.

To be clear, I don't yearn to go back to the good old days when a small group of all-powerful editors unilaterally decided what was important. Too many actually important stories (the genocide in Rwanda, for example) fell through the cracks, while too many actually unimportant ones got front-page coverage. But I also don't think we should jettison that approach altogether. Yahoo News suggests there is some possibility for middle ground: The team combines algorithmic personalization with old-school editorial leadership. Some stories are visible to everyone because they're surpassingly important. Others show up for some users and not others. And while the editorial team at Yahoo spends a lot of time interpreting click data and watching which articles do well and which don't, they're not subservient to it. "Our editors think of the audience as people with interests, as opposed to a flood of directional data," a Yahoo News employee told me. "As much as we love the data, it's being filtered by human beings who are thinking about what the heck it means. Why didn't the article on this topic we think is important for our readers to know about do better? How do we help it find a larger audience?"

And then there are fully algorithmic solutions. For example, why not rely on everyone's idea of what's important? Imagine for a moment that next to each Like button on Facebook was an Important button. You could tag items with one or the other or both. And Facebook could draw on a mix of both signals— what people like, and what they think really matters—to populate and personalize your news feed. You'd have to bet that news about Pakistan would be seen more often—even accounting for everyone's quite subjective definition of what really matters. Collaborative filtering doesn't have to lead to compulsive media: The whole game is in what values the filters seek to pull out. Alternately, Google or Facebook could place a slider bar running from "only stuff I like" to "stuff other people like that I'll probably hate" at the top of search results and the News Feed, allowing users to set their own balance between tight personalization and a more diverse information flow. This approach would have two benefits: It would make clear that there's personalization going on, and it would place it more firmly in the user's control.

There's one more thing the engineers of the filter bubble can do. They can solve for serendipity, by designing filtering systems to expose people to topics outside their normal experience. This will often be in tension with pure optimization in the short term, because a personalization system with an element of randomness will (by definition) get fewer clicks. But as the problems of personalization become better known, it may be a good move in the long run—consumers may choose systems that are good at introducing them to new topics. Perhaps what we need is a kind of anti-Netflix Prize—a Serendipity Prize for systems that are the best at holding readers' attention while introducing them to new topics and ideas.

If this shift toward corporate responsibility seems improbable, it's not without precedent. In the mid-1800s, printing a newspaper was hardly a reputable business. Papers were fiercely partisan and recklessly ideological. They routinely altered facts to suit their owners' vendettas of the day, or just to add color. It was this culture of crass commercialism and manipulation that Walter Lippmann railed against in *Liberty and the News*.

But as newspapers became highly profitable and highly important, they began to change. It became possible, in a few big cities, to run papers that weren't just chasing scandal and sensation—in part, because their owners could afford not to. Courts started to recognize a public interest in journalism and rule accordingly. Consumers started to demand more scrupulous and rigorous editing.

Urged on by Lippmann's writings, an editorial ethic began to take shape. It was never shared universally or followed as well as it could have been. It was always compromised by the business demands of newspapers' owners and shareholders. It failed outright repeatedly—access to power brokers compromised truth telling, and the demands of advertisers overcame the demands of readers. But in the end, it succeeded, somehow, in seeing us through a century of turmoil.

The torch is now being passed to a new generation of curators, and we need them to pick it up and carry it with pride. We need programmers who will build public life and citizenship into the worlds they create. And we need users who will hold them to it when the pressure of monetization pulls them in a different direction.

What Governments and Citizens Can Do

There's plenty that the companies that power the filter bubble can do to mitigate the negative consequences of personalization—the ideas above are just a start. But ultimately, some of these problems are too important to leave in the hands of private actors with profit-seeking motives. That's where governments come in.

Ultimately, as Eric Schmidt told Stephen Colbert, Google is just a company. Even if there are ways of addressing these issues that don't hurt the bottom line—which there may well be—doing so simply isn't always going to be a top-level priority. As a result, after we've each done our part to pop the filter bubble, and after companies have done what they're willing to do, there's probably a need for government oversight to ensure that we control our online tools and not the other way around.

In his book *Republic.com*, Cass Sunstein suggested a kind of "fairness doctrine" for the Internet, in which information aggregators have to expose their audiences to both sides. Though he later changed his mind, the proposal suggests one direction for regulation: Just require curators to behave in a public-oriented way, exposing their readers to diverse lines of argument. I'm skeptical, for some of the same reasons Sunstein abandoned the idea: Curation is a nuanced, dynamic thing, an art as much as a science, and it's hard to imagine how regulating editorial ethics wouldn't inhibit a great deal of experimentation, stylistic diversity, and growth.

As this book goes to press, the U.S. Federal Trade Commission is proposing a Do Not Track list, modeled after the highly successful Do Not Call list. At first blush, it sounds pretty good: It would set up a single place to opt out of the online tracking that fuels personalization. But Do Not Track would probably offer a binary choice—either you're in or you're out—and services that make money on tracking might simply disable themselves for Do Not Track list members. If most of the Internet goes dark for these people, they'll quickly leave the list. And as a result, the process could backfire—"proving" that people don't care about tracking, when in fact what most of us want is more nuanced ways of asserting control.

The best leverage point, in my view, is in requiring companies to give us real control over our personal information. Ironically, although online personalization is relatively new, the principles that ought to support this leverage have been clear for decades. In 1973, the Department of Housing, Education, and Welfare under Nixon recommended that regulation center on what it called Fair Information Practices:

- You should know who has your personal data, what data they have, and how it's used.
- You should be able to prevent information collected about you for one purpose from being used for others.

- You should be able to correct inaccurate information about you.
- Your data should be secure.

Nearly forty years later, the principles are still basically right, and we're still waiting for them to be enforced. We can't wait much longer: In a society with an increasing number of knowledge workers, our personal data and "personal brand" are worth more than they ever have been. Especially if you're a blogger or a writer, if you make funny videos or music, or if you coach or consult for a living, your online data trail is one of your most valuable assets. But while it's illegal to use Brad Pitt's image to sell a watch without his permission, Facebook is free to use your name to sell one to your friends.

In courts around the world, information brokers are pushing this view—"everyone's better off if your online life is owned by us." They argue that the opportunities and control that consumers get by using their free tools outweigh the value of their personal data. But consumers are entirely unequipped to make this calculation—while the control you gain is obvious, the control you lose (because, say, your personal data is used to deny you an opportunity down the road) is invisible. The asymmetry of understanding is vast.

To make matters worse, even if you carefully read a company's privacy policy and decide that giving over rights to your personal information is worth it under those conditions, most companies reserve the right to change the rules of the game at any time. Facebook, for example, promised its users that if they made a connection with a Page, that information would only be shared with their friends. But in 2010, it decided that all of that data should be made fully public; a clause in Facebook's privacy policy (as with many corporate privacy policies) allows it to change the rules *retroactively*. In effect, this gives them nearly unlimited power to dispatch personal data as they see fit.

To enforce Fair Information Practices, we need to start thinking of personal data as a kind of personal property and protecting our rights in it. Personalization is based on an economic transaction in which consumers are at an inherent disadvantage: While Google may know how much your race is worth to Google, you don't. And while the benefits are obvious (free e-mail!), the drawbacks (opportunities and content missed) are invisible. Thinking of personal information as a form of property would help make this a fairer market.

Although personal information is property, it's a special kind of property, because you still have a vested interest in your own data long after it's been exposed. You probably wouldn't want consumers to be able to sell all of their personal data, in perpetuity. France's "moral laws," in which artists retain some control over what's done with a piece after it's been sold, might be a better template. (Speaking of France, while European laws are much closer to Fair Information Practices in protecting personal information, by many accounts the enforcement is much worse, partly because it's much harder for individuals to sue for breaches of the laws.)

Marc Rotenberg, executive director of the Electronic Privacy Information Center, says, "We shouldn't have to accept as a starting point that we can't have free services on the Internet without major privacy violations." And this isn't just about privacy. It's also about how our data shapes the content and opportunities we see and don't see. And it's about being able to track and manage this constellation of data that represents our lives with the same ease that companies like Acxiom and Facebook already do.

Silicon Valley technologists sometimes portray this as an unwinnable fight—people have lost control of their personal data, they'll never regain it, and they just have to grow up and live with it. But legal requirements on personal information need not be foolproof in order to work, any more than legal requirements not to steal are useless because people sometimes still steal things and get away with it. The force of law adds friction to the transmission of some kinds of information—and in many cases, a little friction changes a lot.

And there are laws that do protect personal information even in this day and age. The Fair Credit Reporting Act, for example, ensures that credit agencies have to disclose their credit reports to consumers and notify consumers when they're discriminated against on the basis of reports. That's not much, but given that previously consumers couldn't even see if their credit report contained errors (and 70 percent do, according to U.S. PIRG), it's a step in the right direction.

A bigger step would be putting in place an agency to oversee the use of personal information. The EU and most other industrial nations have this kind of oversight, but the United States has lingered behind, scattering responsibilities for protecting personal information among the Federal Trade Commission, the Commerce Department, and other agencies. As we enter the second decade of the twenty-first century, it's past time to take this concern seriously.

None of this is easy: Private data is a moving target, and the process of balancing consumers and citizens' interests against those of these companies will take a lot of fine-tuning. At worst, new laws could be more onerous than the practices they seek to prevent. But that's an argument for doing this right and doing it soon, before the companies who profit from private information have even greater incentives to try to block it from passing.

Given the money to be made and the power that money holds over the American legislative system, this shift won't be easy. So to rescue our digital environment from itself, we'll ultimately need a new constituency of digital environmentalists—citizens of this new space we're all building who band together to protect what's great about it.

In the next few years, the rules that will govern the next decade or more of online life will be written. And the big online conglomerates are lining up to help write them. The communications giants who own the Internet's physical infrastructure have plenty of political clout. AT&T outranks oil companies and pharmaceutical companies as one of the top four corporate contributors to American politics. Intermediaries like Google get the importance of political influence, too: Eric Schmidt is a frequent White House visitor, and companies like Microsoft, Google, and Yahoo have spent millions on influence-mongering in Washington, D.C. Given all of the Web 2.0 hype about empowerment, it's ironic that the old adage still applies: In the fight for control of the Internet, everyone's organized but the people.

But that's only because most of us aren't in the fight. People who use the Internet and are invested in its future outnumber corporate lobbyists by orders of magnitude. There are literally hundreds of millions of us across all demographics—political, ethnic, socioeconomic, and generational—who have a personal stake in the outcome. And there are plenty of smaller online enterprises that have every interest in ensuring a democratic, public-spirited Web. If the great mass of us decide that an open, public-spirited Internet matters and speak up about it—if we join organizations like Free Press (a nonpartisan grassroots lobby for media reform) and make phone calls to Congress and ask questions at town hall meetings and contribute donations to the representatives who are leading the way—the lobbyists don't stand a chance.

As billions come online in India and Brazil and Africa, the Internet is transforming into a truly global place. Increasingly, it will be the place where we live our lives. But in the end, a small group of American companies may unilaterally dictate how billions of people work, play, communicate, and understand the world. Protecting the early vision of radical connectedness and user control should be an urgent priority for all of us.

ACKNOWLEDGMENTS

Writing may be a lonely profession, but thinking isn't. That's been one of the great gifts of this writing process—the opportunity to think together and learn from some extremely smart and morally thoughtful people. This book wouldn't be the same—and wouldn't be much—without a large team of (sometimes unwitting) collaborators. What follows is my best attempt to credit those who contributed directly. But there's an even larger number whose scholarship or writing or philosophy gave structure to my thoughts or forced me to think in a new way: Larry Lessig, Neil Postman, Cass Sunstein, Marshall McLuhan, Marvin Minsky, and Michael Schudson come to mind as a start. What's good in this book owes a lot to this broad cadre of thinkers. The errors, of course, are all mine.

The Filter Bubble began as a sketched fragment of text jotted down in the first days of 2010. Elyse Cheney, my literary agent, gave me the confidence to see it as a book. Her keen editorial eye, fierce intellect, and refreshingly blunt assessments ("That part's pretty good. This chapter, not so much.") dramatically strengthened the final text. I know it's par for the course to thank one's agent. But Elyse was more than an agent for this book—she was its best proponent and critic, constantly pushing it (and me) to be great. Whether or not the final manuscript met that mark, I've learned a lot, and I'm grateful and deeply appreciative. Her team—Sarah Rainone and Hannah Elnan—were also terrific to work with.

Ann Godoff and Laura Stickney, my editors at Penguin Press, are the other two members of the triumvirate that brought this book into existence. Ann's wisdom helped to shape what this book is about and for whom I've written it; Laura's acute eye and gentle questions and provocations helped me see the gaps, leaps, and snags in the text. I'm indebted to both.

There's another trio that deserves a great deal of credit, not just for getting this book across the finish line in (more or less) one piece, but for inspiring some of the best insights in it. Research assistants Caitlin Petre, Sam Novey, and Julia Kamin scoured the Internet and dug through dusty library books to help me figure out what exactly was going on. Sam, my resident contrarian, constantly pushed me to think more deeply about what I was saying. Julia brought a keen scientific skepticism to the task and protected me from dubious scholarship that I might otherwise have embraced. And Caitlin's great intelligence, hard work, and thoughtful critique were the sources of some of my favorite aha moments. Guys, I couldn't have done it without you. Thank you.

One of the best parts of the writing process was the opportunity to call up or sit down with extraordinary people and ask them questions. I'm thankful to the following folks for responding to my inquiries and helping inform the text: C. W.

Anderson, Ken Auletta, John Battelle, Bill Bishop, Matt Cohler, Gabriella Coleman, Dalton Conley, Chris Coyne, Pam Dixon, Caterina Fake, Matthew Hindman, Bill Joy, Dave Karpf, Jaron Lanier, Steven Levy, Diana Mutz, Nicholas Negroponte, Markus Prior, Robert Putnam, John Rendon, Jay Rosen, Marc Rotenberg, Douglas Rushkoff, Michael Schudson, Daniel Solove, Danny Sullivan, Philip Tetlock, Clive Thompson, and Jonathan Zittrain. Conversations with Ethan Zuckerman, Scott Heiferman, David Kirkpatrick, Clay Shirky, Nicco Mele, Dean Eckles, Jessi Hempel, and Ryan Calo were especially provocative and helpful. Thanks to Nate Tyler and Jonathan McPhie at Google for considering and responding to my inquiries. And strange though it may seem, given my topic, thanks also to my Facebook friends, some real, some virtual, who quickly responded to my queries and were helpful when I was looking for an anecdote or colorful example.

During the writing process, I've received invaluable help from a number of institutions and communities. I don't know where I'd be without the summer months I spent researching and writing at the Blue Mountain Center: many thanks to Ben, Harriet, and my fellow fellows for the space to think, advice (especially from Carey McKenzie), and late-night swims. The Roosevelt Institute was gracious enough to offer a place to hang my hat for the last year: thanks to Andy Rich and Bo Cutter for the intellectual stimulation and great conversations. Micah Sifry and Andrew Raseij, two great friends of online democracy, gave me the space to first make this argument at the Personal Democracy Forum. David Fenton has been there to help with every step of this process, from lending his home for writing and thinking to consulting on the title to helping the book find an audience. David, you're a good friend. And Fenton Communications—especially my kind, thoughtful friend Lisa Witter—generously supported the early investigations that set me on the personalization trail.

There's little I can say to sufficiently thank Team MoveOn, past and present, from whom I've learned so very much about politics, technology, and people. Carrie, Zack, Joan, Patrick, Tom, Nita, Jenn, Ben, Matt, Natalie, Noah, Adam, Roz, Justin, Ilyse, and the whole crew: You're some of the most fiercely thoughtful and inspiring people I've ever met, and I feel lucky to have worked alongside you.

The manuscript was ready to read only weeks before it was due. Wes Boyd, Matt Ewing, Randall Farmer, Daniel Mintz, my parents, Emanuel Pariser and Dora Lievow, and of course Sam, Caitlin, and Julia were all kind enough to put aside busy lives and plow through it. I shudder to think of what might have gone to print without their notes. Todd Rogers, Anne O'Dwyer, Patrick Kane, David Kirkpatrick, and Jessi Hempel were all kind enough to look at pieces of the book, as well. And I can't say thanks enough to Krista Williams and Amanda Katz, whose brilliant editorial thinking helped nurse some somewhat sickly chapters back to health (Krista, a second thanks for your friendship). Stephanie Hopkins and Mirela Iverac provided invaluable last-minute assistance with the manuscript.

I've saved the greatest and most personal debts for last. I've benefited immeasurably in my life from a string of great teachers: Karen Scott, Doug Hamill, and Leslie Simmons at Lincolnville Central School; Jon Potter and Rob Lovell at Camden-Rockport

High; and Barbara Resnik and Peter Cocks at Simon's Rock, among others. Whatever perspicuity I have I owe these folks. And I'm lucky enough to have some truly wonderful friends. I can't list all of you here, but you know who you are. I'm especially grateful for the support and love I've received—in good times and bad—from Aram and Lara Kailian, Tate Hausman, Noah T. Winer, Nick Arons, and Ben and Beth Wikler. It's one of my goals in life to be as good a friend as you've been to me.

My family has also encouraged me and sharpened my thinking every step of the way. Big hugs and even bigger thank-yous to my mother, Dora Lievow, my father and stepmother, Emanuel Pariser and Lea Girardin, and my sister, Ya Jia. Eben Pariser, my brother, not only egged me on but made amazing pizza when I was flagging and helped finalize the manuscript. He's as good a brother as he is a musician (check out his band, Roosevelt Dime, and you'll see how high that compliment is). Bronwen Rice may not be an official family member, but I'll include her here anyway: Bronwen, thanks for keeping me true to myself all of these years.

There are four final people whose generosity, intelligence, and love I appreciate more than I can fully express:

Wes Boyd took a big gamble on a twenty-one-year-old, trusted me more than I trusted myself, and mentored me through eight years at MoveOn. This book draws on many of our conversations over the years—there's no one I enjoy thinking with more. Peter Koechley, my true friend and coconspirator, encouraged me when the going got rough—in the writing process and outside of it. I'm grateful to have a friend who is simultaneously so talented and so decent. Vivien Labaton: I have no sufficient superlatives, so I'll just go with the colloquial. You're the best. And finally, there's Gena Konstantinakos. Gena, you've borne the brunt of this project more than anyone—the months of working weekends and late nights and early mornings at the office, the stress during revisions, and the constantly extended deadlines. You took it in stride and then some, giving me pep talks, helping me sort out chapters on note cards, and cheering me on all the way. I'm still amazed, some days, to wake up with someone so smart, beautiful, talented, principled, and good-spirited in my life. I love you.

FURTHER READING

Alexander, Christopher, Sara Ishikawa, and Murray Silverstein. *A Pattern Language: Towns, Buildings, Construction*. New York: Oxford University Press, 1977.

Anderson, Benedict. *Imagined Communities: Reflections on the Origin and Spread of Nationalism*. New York: Verso, 1991.

Battelle, John. *The Search: How Google and Its Rivals Rewrote the Rules of Business and Transformed Our Culture*. New York: Portfolio, 2005.

Berger, John. *Ways of Seeing*. New York: Penguin, 1973.

Bishop, Bill. *The Big Sort: Why the Clustering of Like-Minded America Is Tearing Us Apart*. New York: Houghton Mifflin Company, 2008.

Bohm, David. *On Dialogue*. New York: Routledge, 1996.

Conley, Dalton. *Elsewhere, U.S.A.: How We Got from the Company Man, Family Dinners, and the Affluent Society to the Home Office, BlackBerry Moms, and Economic Anxiety*. New York: Pantheon Books, 2008.

Dewey, John. *Public and Its Problems*. Athens, OH: Swallow Press, 1927.

Heuer, Richards J. *Psychology of Intelligence Analysis*. Washington, D.C.: Central Intelligence Agency, 1999.

Inglehart, Ronald. *Modernization and Postmodernization*. Princeton: Princeton University Press, 1997.

Kelly, Kevin. *What Technology Wants*. New York: Viking, 2010.

Koestler, Arthur. *The Act of Creation*. New York: Arkana, 1989.

Lanier, Jaron. *You Are Not a Gadget: A Manifesto*. New York: Alfred A. Knopf, 2010.

Lessig, Lawrence. *Code: And Other Laws of Cyberspace, Version 2.0*. New York: Basic Books, 2006.

Lippmann, Walter. *Liberty and the News*. Princeton: Princeton University Press, 1920.

Minsky, Marvin. *A Society of Mind*. New York: Simon and Schuster, 1988.

Norman, Donald A. *The Design of Everyday Things*. New York: Basic Books, 1988.

Postman, Neil. *Amusing Ourselves to Death: Public Discourse in the Age of Show Business*. New York: Penguin Books, 1985.

Schudson, Michael. *Discovering the News: A Social History of American Newspapers*. New York: Basic Books, 1978.

Shields, David. *Reality Hunger: A Manifesto*. New York: Alfred A. Knopf, 2010.

Shirky, Clay. *Here Comes Everybody: The Power of Organizing Without Organizations*. New York: The Penguin Press, 2008.

Solove, Daniel J. *Understanding Privacy*. Cambridge, MA: Harvard University Press, 2008.

Sunstein, Cass R. *Republic.com 2.0*. Princeton: Princeton University Press, 2007.

Turner, Fred. *From Counterculture to Cyberculture: Stewart Brand, the Whole Earth Network, and the Rise of Digital Utopianism*. Chicago: The University of Chicago Press, 2006.

Watts, Duncan J. *Six Degrees: The Science of a Connected Age*. New York: W.W. Norton & Company, 2003.

Wu, Tim. *The Master Switch: The Rise and Fall of Information Empires*. New York: Alfred A. Knopf, 2010.

Zittrain, Jonathan. *The Future of the Internet—And How to Stop It*. New Haven: Yale University Press, 2008.

Introduction

1 **"A squirrel dying":** David Kirkpatrick, *The Facebook Effect: The Inside Story of the Company That Is Connecting the World* (New York: Simon and Schuster, 2010), 296.

1 **"thereafter our tools shape us":** Marshall McLuhan, *Understanding Media: The Extensions of Man* (Cambridge: MIT Press, 1994).

1 **"Personalized search for everyone":** *Google Blog*, Dec. 4, 2009, accessed Dec. 19, 2010, http://googleblog.blogspot.com/2009/12/personalized -search-for-everyone.html.

2 **Google would use fifty-seven *signals*:** Author interview with confidential source.

6 ***Wall Street Journal* study:** Julia Angwin, "The Web's New Gold Mine: Your Secrets," *Wall Street Journal*, July 30, 2010, accessed Dec. 19, 2010, http://online.wsj.com/article/SB10001424052748703940904575395073512989404.html.

6 **"Yahoo":** Although the official trademark is Yahoo!, I've omitted the exclamation point throughout this book for easier reading.

6 **site installs 223 tracking cookies:** Angwin, "The World's New Gold Mine," July 30, 2010.

6 **Teflon-coated pots:** At the time of writing, ABC News used a piece of sharing software called "AddThis." When you use AddThis to share a piece of content on ABC News's site (or anyone else's), AddThis

places a tracking cookie on your computer that can be used to target advertising to people who share items from particular sites.

6   "the cost is information about you": Chris Palmer, phone interview with author, Dec 10, 2010.

7   accumulated an average of 1,500 pieces of data: Stephanie Clifford, "Ads Follow Web Users, and Get More Personal," *New York Times*, July 30, 2009, accessed Dec. 19, 2010, www.nytimes.com/2009/07/31/business/media/31privacy.html.

7   96 percent of Americans: Richard Behar, "Never Heard of Acxiom? Chances Are It's Heard of You." *Fortune*, Feb. 23, 2004, accessed Dec. 19, 2010, http://money.cnn.com/magazines/fortune/fortune_archive/2004/02/23/362182/index.htm.

8   Netflix can predict: Marshall Kirkpatrick, "They Did It! One Team Reports Success in the $1m Netflix Prize," *ReadWriteWeb*, June 26, 2009, accessed Dec. 19, 2010, www.readwriteweb.com/archives/they_did_it_one_team_reports_success_in_the_1m_net.php.

8   Web site that isn't customized . . . will seem quaint: Marshall Kirpatrick, "Facebook Exec: All Media Will Be Personalized in 3 to 5 Years," *ReadWriteWeb*, Sept. 29, 2010, accessed Jan. 30, 2011, www.readwriteweb.com/archives/facebook_exec_all_media_will_be_personalized_in_3.php.

8   "now the web is about 'me'": Josh Catone, "Yahoo: The Web's Future Is Not in Search," *ReadWriteWeb*, June 4, 2007, accessed Dec. 19, 2010, www.readwriteweb.com/archives/yahoo_personalization.php.

8   "tell them what they should be doing": James Farrar, "Google to End Serendipity (by Creating It)," *ZDNet*, Aug. 17, 2010, accessed Dec. 19, 2010, www.zdnet.com/blog/sustainability/google-to-end-serendipity-by-creating-it/1304.

8   are becoming a primary news source: Pew Research Center, "Americans Spend More Time Following the News," Sept. 12, 2010, accessed Feb 7, 2011, http://people-press.org/report/?pageid=1793.

8   million more people joining each day: Justin Smith, "Facebook Now Growing by Over 700,000 Users a Day, and New Engagement Stats," July 2, 2009, accessed Feb. 7, 2011, www.insidefacebook.com/2009/07/02/

facebook-now-growing-by-over-700000-users-a-day-updated-engage
ment-stats/.

8   **biggest source of news in the world:** Ellen McGirt, "Hacker. Drop
     out. CEO," *Fast Company*, May 1, 2007, accessed Feb. 7, 2011, www
     .fastcompany.com/magazine/115/open_features-hacker-dropout-ceo
     .html.

11   **information: 900,000 blog posts, 50 million tweets:** "Measuring
     tweets," *Twitter* blog, Feb. 22, 2010, accessed Dec. 19, 2010, http://
     blog.twitter.com/2010/02/measuring-tweets.html.

11   **60 million Facebook status updates, and 210 billion e-mails:** "A Day
     in the Internet," Online Education, accessed Dec. 19, 2010, www
     .onlineeducation.net/internet.

11   **about 5 billion gigabytes:** M. G. Siegler, "Eric Schmidt: Every 2 Days
     We Create as Much Information as We Did up to 2003," *TechCrunch*
     blog, Aug. 4, 2010, accessed Dec. 19, 2010, http://techcrunch.com/
     2010/08/04/schmidt-data.

11   **two new stadium-size complexes:** Paul Foy, "Gov't Whittles Bidders for
     NSA's Utah Data Center," Associated Press, Apr. 21, 2010, accessed
     Dec. 19, 2010, http://abcnews.go.com/Business/wireStory?id=10438827
     &page=2.

11   **new units of measurements:** James Bamford, "Who's in Big Brother's
     Database?," *The New York Review of Books*, Nov 5, 2009, accessed Feb. 8,
     2011,www.nybooks.com/articles/archives/2009/nov/o5/whos-in-big
     -brothers-database.

11   **the attention crash:** Steve Rubel, "Three Ways to Mitigate the Atten-
     tion Crash, Yet Still Feel Informed," *Micro Persuasion* (Steve Rubel's
     blog), Apr. 30, 2008, accessed Dec. 19, 2010, www.micropersuasion
     .com/2008/04/three-ways-to-m.html.

13   **"back in the bottle":** Danny Sullivan, phone interview with author,
     Sept 10, 2010.

13   **part of our daily experience:** Cass Sunstein, *Republic.com 2.0.* (Princeton:
     Princeton University Press, 2007).

13–14 **"skew your perception of the world":** Ryan Calo, phone interview
     with author, Dec. 13, 2010.

14   **"the psychological equivalent of obesity":** danah boyd, "Streams of
     Content, Limited Attention: The Flow of Information through Social
     Media," speech, Web 2.0 Expo. (New York: 2009), accessed July 19,
     2010, www.danah.org/papers/talks/Web2Expo.html.

15   **"strategically time" their online solicitations:** "Ovulation Hormones
     Make Women 'Choose Clingy Clothes,'" BBC News, Aug. 5, 2010,
     accessed Feb: 8, 2011, www.bbc.co.uk/news/health-10878750.

16   **third-party marketing firms:** "Preliminary FTC Staff Privacy Report,"
     remarks of Chairman Jon Leibowitz, as prepared for delivery, Dec. 1,
     2010, accessed Feb. 8, 2011, www.ftc.gov/speeches/leibowitz/101201
     privacyreportremarks.pdf.

16   **Yochai Bentler argues:** Yochai Benkler, "Siren Songs and Amish Chil-
     dren: Autonomy, Information, and Law," *New York University Law
     Review*, Apr. 2001.

17   **tap into lots of different networks:** Robert Putnam, *Bowling Alone: The
     Collapse and Revival of American Community* (New York: Simon and
     Schuster, 2000).

17   **"make us all next door neighbors":** Thomas Friedman, "It's a Flat
     World, After All," *New York Times*, Apr. 3, 2005, accessed Dec. 19, 2010,
     www.nytimes.com/2005/04/03/magazine/03DOMINANCE.html?
     pagewanted=all.

17   **"smaller and smaller and faster and faster":** Thomas Friedman, *The
     Lexus and the Olive Tree* (New York: Random House, 2000), 141.

18   **"closes the loop on pecuniary self-interest":** Clive Thompson, inter-
     view with author, Brooklyn, NY, Aug. 13, 2010.

18   **"Customers are always right, but people aren't":** Lee Siegel, *Against
     the Machine: Being Human in the Age of the Electronic Mob* (New York:
     Spiegel and Grau, 2008), 161.

18   **thirty-six hours a week watching TV:** "Americans Using TV and Inter-
     net Together 35% More Than A Year Ago," Nielsen Wire, Mar. 22,
     2010, accessed Dec. 19, 2010, http://blog.nielsen.com/nielsenwire/
     online_mobile/three-screen-report-q409.

19   **"civilization of Mind in cyberspace":** John Perry Barlow, "A Cyberspace
     Independence Declaration," Feb. 9, 1996, accessed Dec. 19, 2010,

http://w2.eff.org/Censorship/Internet_censorship_bills/barlow_0296
.declaration.

19   "code is law": Lawrence Lessig, *Code 2.0* (New York: Basic Books, 2006), 5.

Chapter One: The Race for Relevance

21   "If you're not paying for something": *MetaFilter* blog, accessed Dec. 10,
       2010, www.metafilter.com/95152/Userdriven-discontent.

22   "vary sex, violence, and political leaning": Nicholas Negroponte, *Being
       Digital* (New York: Knopf, 1995), 46.

22   "the Daily Me": Ibid., 151.

22   "Intelligent agents are the unequivocal future": Negroponte, Mar. 1,
       1995, e-mail to the editor, Wired.com, Mar. 3, 1995, www.wired.com/
       wired/archive/3.03/negroponte.html.

23   "The agent question looms": Jaron Lanier, "Agents of Alienation,"
       accessed Jan. 30, 2011, www.jaronlanier.com/agentalien.html

24   twenty-five worst tech products: Dan Tynan, "The 25 Worst Tech Prod-
       ucts of All Time," *PC World*, May 26, 2006, accessed Dec. 10, 2010,
       www.pcworld.com/article/125772-3/the_25_worst_tech_products
       _of_all_time.html#bob.

24   invested over $100 million: Dawn Kawamoto, "Newsmaker: Riding the
       next technology wave," CNET News, Oct. 2, 2003, accessed Jan. 30,
       2011, http://news.cnet.com/2008-7351-5085423.html.

25   "he's a lot like John Irving": Robert Spector, *Get Big Fast* (New York:
       HarperBusiness, 2000), 142.

25   "small Artificial Intelligence company": Ibid., 145.

26   surprised to find them at the top: Ibid., 27.

26   Random House, controlled only 10 percent: Ibid., 25.

26   so many of them—3 million active titles: Ibid., 25.

27   They called their field "cybernetics": Barnabas D. Johnson, "Cybernet-
       ics of Society," The Jurlandia Institute, accessed Jan. 30, 2011, www
       .jurlandia.org/cybsoc.htm.

27   PARC was known for: Michael Singer, "Google Gobbles Up Outride,"
       InternetNews.com, Sept. 21, 2001, accessed Dec. 10, 2010, www

.internetnews.com/bus-news/article.php/889381/Google-Gobbles -Up-Outride.html.

27   **collaborative filtering:** Moya K. Mason, "Short History of Collaborative Filtering," accessed Dec. 10, 2010, www.moyak.com/papers/collaborative-filtering.html.

28   **"handle any incoming stream of electronic documents":** David Goldberg, David Nichols, Brian M. Oki, and Douglas Terry, "Using Collaborative Filtering to Weave an Information Tapestry," *Communications of the ACM* 35 (1992), 12: 61.

28   **"sends replies as necessary":** Upendra Shardanand, "Social Information Filtering for Music Recommendation" (graduate diss., Massachusetts Institute of Technology, 1994).

29   **fewer health books:** Martin Kaste, "Is Your E-Book Reading Up On You?," NPR.org, Dec. 15, 2010, accessed Feb. 8, 2010, www.npr.org /2010/12/15/132058735/is-your-e-book-reading-up-on-you.

30   **as if by an "objective" recommendation:** Aaron Shepard, *Aiming at Amazon: The NEW Business of Self Publishing, Or How to Publish Your Books with Print on Demand and Online Book Marketing* (Friday Harbor, WA: Shepard Publications, 2006), 127.

30   **"notion of 'relevant'":** Sergey Brin and Lawrence Page, "The Anatomy of a Large-Scale Hypertextual Web Search Engine," Section 1.3.1.

31   **"advertising causes enough mixed incentives":** Ibid., Section 8, Appendix A.

32   **"very difficult to get this data":** Ibid., Section 1.3.2.

33   **black-ops kind of feel:** Saul Hansell, "Google Keeps Tweaking Its Search Engine," *New York Times*, June 3, 2007, accessed Feb. 7, 2011, www.nytimes.com/2007/06/03/business/yourmoney/03google .html?_r=1.

33   **"give back exactly what you want":** David A. Vise and Mark Malseed, *The Google Story* (New York: Bantam Dell, 2005), 289.

34   **"ancient shark teeth":** Patent full text, accessed Dec. 10, 2010, http:// patft.uspto.gov/netacgi/nph-Parser?Sect1=PTO2&Sect2=HITOFF&u =%2Fnetahtml%2FPTO%2Fsearch-adv.htm&r=1&p=1&f=G&l=50&d =PTXT&S1=7,451,130.PN.&OS=pn/7,451,130&RS=PN/7,451,13,

35   "could call that artificial intelligence": Lawrence Page, Google Zeitgeist Europe Conference, May 2006.

35   "answer a more hypothetical question": BBC News, "Hyper-personal Search 'Possible,'" June 20, 2007, accessed Dec. 10, 2010, http://news .bbc.co.uk/2/hi/technology/6221256.stm.

36   "We're a utility": David Kirkpatrick, "Facebook Effect," *New York Times*, June 8, 2010, accessed Dec. 10, 2010, www.nytimes.com/2010/06/08/ books/excerpt-facebook-effect.html?pagewanted=1.

37   "more news in a single day": Ellen McGirt, "Hacker. Dropout. CEO," *Fast Company*, May 1, 2007, accessed Feb. 7, 2011, http://www.fast company.com/magazine/115/open_features-hacker-dropout-ceo .html.

37   it rests on three factors: Jason Kincaid, "EdgeRank: The Secret Sauce That Makes Facebook's News Feed Tick," *TechCrunch* blog, Apr. 22, 2010, accessed Dec. 10, 2010, http://techcrunch.com/2010/04/22/ facebook-edgerank.

38   the 300 million user mark: Mark Zuckerberg, "300 Million and On," *Facebook* blog, Sept. 15, 2009, accessed Dec. 10, 2010, http://blog .facebook.com/blog.php?post=136782277130.

38   the *Washington Post* homepage: Full disclosure: In the spring of 2010, I briefly consulted with the *Post* about its online communities and Web presence.

39   "the most transformative thing": Caroline McCarthy, "Facebook F8: One Graph to Rule Them All," CNET News, Apr. 21, 2010, accessed Dec. 10, 2010, http://news.cnet.com/8301-13577_3-20003053-36 .html.

39   sharing 25 billion items a month: M. G. Siegler, "Facebook: We'll Serve 1 Billion Likes on the Web in Just 24 Hours," *TechCrunch* blog, Apr. 21, 2010, accessed Dec. 10, 2010, http://techcrunch.com/2010/04/21/ facebook-like-button.

42   Acxiom knew more: Richard Behar, "Never Heard of Acxiom? Chances Are It's Heard of You," *Fortune*, Feb. 23, 2004, accessed Dec. 10, 2010, http://money.cnn.com/magazines/fortune/fortune_archive/2004/ 02/23/362182/index.htm.

43   **serves most of the largest companies in America:** InternetNews.com
      Staff, "Acxiom Hacked, Customer Information Exposed," *InternetNews*
      .com, Aug. 8, 2003, accessed Dec. 10, 2010, www.esecurityplanet.com/
      trends/article.php/2246461/Acxiom-Hacked-Customer-Information
      -Exposed.htm.

43   **"product we make is data":** Behar, "Never Heard of Acxiom?"

44   **auctions it off to the company with the highest bid:** Stephanie Clif-
      ford, "Your Online Clicks Have Value, for Someone Who Has Some-
      thing to Sell," *New York Times*, Mar. 25, 2009, accessed Dec. 10, 2010,
      www.nytimes.com/2009/03/26/business/media/26adco.html?_r=2.

44   **takes under a second:** The Center for Digital Democracy, U.S. Public
      Interest Research Group, and the World Privacy Forum's complaint to
      the Federal Trade Commission, Apr. 8, 2010, accessed Dec. 10, 2010,
      http://democraticmedia.org/real-time-targeting.

44   **leave without buying anything:** Press release, FetchBack Inc., Apr. 13,
      2010, accessed Dec. 10, 2010, www.fetchback.com/press_041310.html.

45   **"62 billion real-time attributes a year":** Center for Digital Democracy,
      U.S. PIRG, and the World Privacy Forum's complaint.

45   **the Rubicon Project:** Ibid.

Chapter Two: The User Is the Content

47   **"undermines the democratic way of life":** John Dewey, *Essays, Reviews,*
      and Miscellany, 1939–1941, The Later Works of John Dewey, 1925–1953,
      vol.14 (Carbondale: Southern Illinois University Press, 1998), 227.

47   **"been tailored for them":** Holman W. Jenkins Jr., "Google and the
      Search for the Future," *Wall Street Journal*, Aug. 14, 2010, accessed
      Dec. 11, 2010, http://online.wsj.com/article/SB10001424052748704
      901104575423294099527212.html.

48   **"don't know which half":** John Wanamaker, U.S. department store
      merchant, as quoted in Marilyn Ross and Sue Collier, *The Complete*
      Guide to Self-Publishing (Cincinnati: Writer's Digest Books, 2010), 344.

49   **One executive in the marketing session:** I wasn't able to identify him
      in my notes.

49   **Now, in 2010, they only received:** Interactive Advertising Bureau PowerPoint, report, "Brand Advertising Online and The Next Wave of M&A," Feb. 2010.

50   **target premium audiences in "other, cheaper places":** Ibid.

50   **"denied an assured access to the facts":** Walter Lippmann, *Liberty and the News* (Princeton: Princeton University Press, 1920), 6.

50   **blogs remain incredibly reliant on them:** Pew Research Center, "How Blogs and Social Media Agendas Relate and Differ from the Traditional Press," May 23, 2010, accessed Dec. 11, 2010, www.journalism.org/node/20621.

52–53   **"these documents are forgeries":** Peter Wallsten, "'Buckhead,' Who Said CBS Memos Were Forged, Is a GOP-Linked Attorney," *Los Angeles Times*, Sept. 18, 2004, accessed Dec. 11, 2010, http://seattletimes .nwsource.com/html/nationworld/2002039080_buckhead18.html.

53   **"We should not have used them":** Associated Press, "CBS News Admits Bush Documents Can't Be Verified," Sept. 21, 2004, accessed Dec. 11, 2010, www.msnbc.msn.com/id/6055248/ns/politics.

54   **paying attention to the story:** *The Gallup Poll: Public Opinion 2004* (Lanham, MD: Rowman & Littlefield, 2006), http://books.google.com/books?id=uqqp-sDCjo4C&pg=PA392&lpg=PA392&dq=public+opinion +poll+on+dan+rather+controversy&source=bl&ots=CPGu03cpsn &sig=9XT-li8ar2GOXxfVQWCcGNHIxTg&hl=en&ei=uw_7TLK9O MGB8gb3r72ACw&sa=X&oi=book_result&ct=result&resnum=1&ved =0CBcQ6AEwAA#v=onepage&q=public%20opinion%20poll%20 on%20dan%20rather%20controversy&f=true.

54   **"a crisis in journalism":** Lippmann, *Liberty and the News*, 64.

56   **at this point that newspapers came to carry:** This section was informed by the wonderful Michael Schudson, *Discovering the News: A Social History of American Newspapers* (New York: Basic Books, 1978).

57   **"They goose-stepped it":** Lippmann, *Liberty and the News*, 4.

57   **"what [the average citizen] shall know":** Ibid., 7.

58   **"distinctive member of a community":** John Dewey, *Essays, Reviews, and Miscellany, 1939–1941, The Later Works of John Dewey, 1925–1953*, vol. 2 (Carbondale: Southern Illinois University Press, 1984), 332.

59 **calls the 2000s the disintermediation decade:** Jon Pareles, "A World of Megabeats and Megabytes," *New York Times*, Dec. 30, 2009, accessed Dec. 11, 2010, www.nytimes.com/2010/01/03/arts/music/03tech.html.

59 *Disintermediation*—**the elimination of middlemen:** Dave Winer, Dec. 7, 2005, Dave Winer's blog, *Scripting News*, accessed Dec. 11, 2010, http://scripting.com/2005/12/07.html#.

59 **"It sucks power out of the center":** Esther Dyson, "Does Google Violate Its 'Don't Be Evil' Motto?," *Intelligence Squared US.* Debate between Esther Dyson, Siva Vaidhyanathan, Harry Lewis, Randal C. Picker, Jim Harper, and Jeff Jarvis (New York, NY) Nov. 18, 2008, accessed Feb. 7, 2011, www.npr.org/templates/story/story.php?storyId=97216369.

60 **the Latin for "middle layer":** Hat tip to Clay Shirky for introducing me to this fact in his conversation with Jay Rosen. Clay Shirky interview by Jay Rosen, video, chap. 5 "Why Study Media?" *NYU Primary Sources* (New York, NY), 2011, accessed Feb 9, 2011, http://nyuprimarysources.org/video-library/jay-rosen-and-clay-shirky/.

61 **"many wresting power from the few":** Lev Grossman, "Time's Person of the Year: You," *Time*, Dec. 13, 2006, accessed Dec. 11, 2010, www.time.com/time/magazine/article/0,9171,1569514,00.html.

61 **"did not eliminate intermediaries":** Jack Goldsmith and Tim Wu, *Who Controls the Internet? Illusions of a Borderless World* (New York: Oxford University Press, 2006), 70.

62 **"It will remember what you know":** Danny Sullivan, "Google CEO Eric Schmidt on Newspapers & Journalism," Search Engine Land, Oct. 3, 2009, accessed Dec. 11, 2010, http://searchengineland.com/google-ceo-eric-schmidt-on-newspapers-journalism-27172.

62 **"bringing the content to the right group":** "Krishna Bharat Discusses the Past and Future of Google News," *Google News* blog, June 15, 2010, accessed Dec. 11, 2010, http://googlenewsblog.blogspot.com/2010/06/krishna-bharat-discusses-past-and.html.

62 **"We pay attention":** Ibid.

63 **"most important, their social circle":** Ibid.

63 **"make it available to publishers":** Ibid.

63   **Americans lost more faith in news:** "Press Accuracy Rating Hits Two
     Decade Low; Public Evaluations of the News Media: 1985–2009," Pew
     Research Center for the People and the Press, Sept. 13, 2009, accessed
     Dec. 11, 2010, http://people-press.org/report/543/.

64   ***New York Times* and some random blogger":** Author's interview with
     Yahoo News executive. Sept. 22, 2010. This interview was conducted
     in confidence.

65   **unplugging from cable TV offerings:** Erick Schonfeld, "Estimate:
     800,000 U.S. Households Abandoned Their TVs for the Web,"
     TechCrunch blog, Apr. 13, 2010, accessed Dec. 11, 2010, http://
     techcrunch.com/2010/04/13/800000-households-abandoned-tvs-web;
     "Cable TV Taking It on the Chin," www.freemoneyfinance.com/2010/11/
     cable-tv-taking-it-on-the-chin.html; and Peter Svensson, "Cable Sub-
     scribers Flee, but Is Internet to Blame?" http://finance.yahoo.com/news/
     Cable-subscribers-flee-but-is-apf-3875814716.html?x=0.

65   **"change the ad industry forever":** "Google Vice President: Online
     Video and TV Will Converge," June 25, 2010, Appmarket.tv, accessed
     Dec. 11, 2010, www.appmarket.tv/news/160-breaking-news/440-google
     -vice-president-online-video-and-tv-will-converge.html.

66   **know people who live near us:** Bill Bishop, *The Big Sort: Why the Clus-
     tering of Like-Minded America Is Tearing Us Apart* (New York: Hough-
     ton Mifflin, 2008), 35.

67   **"watch television to turn your brain off":** Jason Snell, "Steve Jobs on the
     Mac's 20th Anniversary," *Macworld*, Feb. 2, 2004, accessed Dec. 11, 2010,
     www.macworld.com/article/29181/2004/02/themacturns20jobs.html.

67   **thirty-six hours a week:** "Americans Using TV and Internet Together
     35% More Than a Year Ago," nielsenwire, Mar. 22, 2010, accessed
     Dec. 11, 2010, http://blog.nielsen.com/nielsenwire/online_mobile/three
     -screen-report-q409.

68   **quit channel surfing far more quickly:** Paul Klein, as quoted in Marcus
     Prior, *Post-Broadcast Democracy* (New York: Cambridge University
     Press, 2007), 39.

68   **like your own personal TV channel:** "YouTube Leanback Offers Effort-
     less Viewing," *YouTube* blog, July 7, 2010, accessed Dec. 11, 2010,

http://youtube-global.blogspot.com/2010/07/youtube-leanback-offers
-effortless.html.

69   **onto the Big Board, and you're liable to get a raise:** Ben McGrath,
"Search and Destroy: Nick Denton's Blog Empire," *New Yorker*, Oct.
18, 2010, accessed Dec. 11, 2010, www.newyorker.com/reporting/
2010/10/18/101018fa_fact_mcgrath?currentPage=all.

70   **"come to us for our judgment":** Jeremy Peters, "Some Newspapers,
Tracking Readers Online, Shift Coverage," *New York Times*, Sept. 5,
2010, accessed Dec. 11, 2010, www.nytimes.com/2010/09/06/
business/media/06track.html.

71   **gin up stories that will get clicks:** Danna Harman, "In Chile, Instant
Web Feedback Creates the Next Day's Paper," *Christian Science Monitor*, Dec. 1, 2004, accessed Dec. 11, 2010, www.csmonitor.com/2004/
1201/p01s04-woam.html.

71   **"creating content in response to audience insight":** Jeremy Peters, "At
Yahoo, Using Searches to Steer News Coverage," *New York Times*, July
5, 2010, accessed Dec. 11, 2010, www.nytimes.com/2010/07/05/
business/media/05yahoo.html.

72   **the newspaper's most e-mailed stories:** Jonah A. Berger and Katherine
L. Milkman, "Social Transmission and Viral Culture," Social Science
Research Network Working Paper Series (Dec. 25, 2009): 2.

72   **"Woman in Sumo Wrestler Suit":** *Huffington Post*, "The Craziest Headline Ever," June 23, 2010, accessed Dec. 11, 2010, www.huffingtonpost
.com/2010/06/23/craziest-bar-ever-discove_n_623447.html.

72   **sex with a horse:** Danny Westneat, "Horse Sex Story Was Online Hit,"
Seattle Times, Dec. 30, 2005, accessed Dec. 11, 2010, http://seattletimes
.nwsource.com/html/localnews/2002711400_danny30.html.

72   **world's ugliest dog:** Ben Margot, "Rescued Chihuahua Princess Abby
Wins World's Ugliest Dog Contest, Besting Boxer Mix Pabst," *Los
Angeles Times*, June 27, 2010, accessed Dec. 11, 2010, http://latimes
blogs.latimes.com/unleashed/2010/06/rescued-chihuahua-princess
-abby-wins-worlds-ugliest-dog-contest-besting-boxer-mix-pabst.html.

72   **"everyone sees the same thing":** Carl Bialik, "Look at This Article. It's
One of Our Most Popular," *Wall Street Journal*, May 20, 2009.

73   "little need to share marketing information": Andrew Alexander,
      "Making the Online Customer King at The Post," *Washington Post*, July
      11, 2010, accessed Dec. 11, 2010, www.washingtonpost.com/wp-dyn/
      content/article/2010/07/09/AR2010070903802.html.

73   "whether you want to hear this or not": Nicholas Negroponte, inter-
      view with author, Truckee, CA, Aug. 5, 2010.

73   "Gawker's Big Board is a scary extreme": Professor Michael Schudson,
      interview with author, New York, NY, Aug. 13, 2010.

73   stories about the war in Afghanistan: Simon Dumenco, "Google News
      Cares More About Facebook, Twitter and Apple Than Iraq, Afghani-
      stan," *Advertising Age*, June 23, 2010, accessed Feb. 9, 2011, http://
      adage.com/mediaworks/article?article_id=144624.

74   "not to pursue some important stories": Alexander, "Making the
      Online Customer King."

75   "periodically be alarmed when there is a crisis?": Shirky, interviewed
      by Jay Rosen.

75   "consequences of conjoint and interacting behavior": John Dewey, *The
      Public and Its Problems* (Athens, OH: Swallow Press, 1927), 126.

Chapter Three: The Adderall Society

77   "contact with persons dissimilar to themselves": John Stuart Mill, *The
      Principles of Political Economy* (Amherst, MA: Prometheus Books,
      2004), 543.

77   "reminds one more of a sleepwalker's": Arthur Koestler, *The Sleepwalk-
      ers: A History of Man's Changing Vision of the Universe* (New York:
      Penguin, 1964), 11.

78   "but I don't want to talk here": Henry Precht, interview with Ambassa-
      dor David E. Mark, Foreign Affairs Oral History Project, Association for
      Diplomatic Studies and Training, July 28, 1989, accessed Dec. 14, 2010,
      http://memory.loc.gov/service/mss/mssmisc/mfdip/2005%20txt%
      20files/2004mar02.txt.

78   the two men planned a meeting: Ibid.

78   "all I want is my money": Ibid.

78   "I was snookered": John Limond Hart, *The CIA's Russians* (Annapolis: Naval Institute Press, 2003), 132.

78   defect and resettle in the United States: Ibid., 135.

79   James Jesus Angleton . . . was skeptical: Ibid., 140.

79   CIA's documents indicated otherwise: "Yuri Ivanovich Nosenko, a Soviet defector, Died on August 23rd, Aged 80," *Economist*, Sept. 4, 2008, accessed Dec. 14, 2010, www.economist.com/node/12051491.

79   subjected to polygraph tests: Ibid.

80   sent to the Russian front as punishment: Richards J. Heuer Jr., "Nosenko: Five Paths to Judgment," *Studies in Intelligence* 31, no. 3 (Fall 1987).

80   set him up in a new identity: David Stout, "Yuri Nosenko, Soviet Spy Who Defected, Dies at 81," *New York Times*, Aug. 27, 2008, accessed Dec. 14, 2010, www.nytimes.com/2008/08/28/us/28nosenko .html?scp=1&sq=nosenko&st=cse.

80   news of his death was relayed: Ibid.

81   full of laudatory comments: Richards J. Heuer Jr., *Psychology of Intelligence Analysis* (Alexandria, VA: Central Intelligence Agency, 1999).

81   "analysts should be self-conscious": Ibid., xiii.

82   secondhand and in a distorted form: Ibid., xx–xxi.

82   "To achieve the clearest possible image": Ibid., xxi–xxii.

83   "predictably irrational": Dan Ariely, *Predictably Irrational: The Hidden Forces That Shape Our Decisions* (New York: HarperCollins, 2008)

83   figuring out what makes us happy: Dan Gilbert, *Stumbling on Happiness* (New York: Knopf, 2006).

83   only one part of the story: Kathryn Schulz, *Being Wrong: Adventures in the Margin of Error* (New York: HarperCollins, 2010).

84   "Information wants to be reduced": Nassim Nicholas Taleb, *The Black Swan: The Impact of the Highly Improbable* (New York: Random House, 2007), 64.

85   quickly converted into schemata: Doris Graber, *Processing the News: How People Tame the Information Tide* (New York: Longman, 1988).

85   "condensation of all features of a story": Ibid., 161.

85   **woman celebrating her birthday:** Steven James Breckler, James M. Olson, and Elizabeth Corinne Wiggins, *Social Psychology Alive* (Belmont, CA: Thomson Wadsworth, 2006), 69.

86   **added details to their memories:** Graber, *Processing the News*, 170.

86   **Princeton versus Dartmouth:** A. H. Hastorf and H. Cantril, "They Saw a Game: A Case Study," *Journal of Abnormal and Social Psychology* 49: 129–34.

87   **experts' predictions weren't even close:** Philip E. Tetlock, *Expert Political Judgment: How Good Is It? How Can We Know?* (Princeton: Princeton University Press, 2005).

88   **a process of assimilation and accommodation:** Jean Piaget, *The Psychology of Intelligence* (New York: Routledge & Kegan Paul, 1950).

89   **the idea that Obama was a Muslim:** Jonathan Chait, "How Republicans Learn That Obama Is Muslim, *New Republic*, Aug. 27, 2010, www.tnr.com/blog/jonathan-chait/77260/how-republicans-learn -obama-muslim.

89   **"actually become mis-educated":** Ibid.

89   **two modified versions of "The Country Doctor":** Travis Proulx and Steven J. Heine, "Connections from Kafka: Exposure to Meaning Threats Improves Implicit Learning of an Artificial Grammar," *Psychological Science* 20, no. 9 (2009): 1125–31.

90   **"A severe snowstorm filled the space":** Franz Kafka, *A Country Doctor* (Prague: Twisted Spoon Press, 1997).

90   **"Once one responds to a false alarm":** Ibid.

90   **"strived to make sense":** Proulx and Heine, "Connections from Kafka."

91   **presented with an "information gap":** George Loewenstein, "The Psychology of Curiosity: A Review and Reinterpretation," *Psychological Bulletin* 116, no. 1 (1994): 75–98, https://docs.google.com/viewer?url= www.andrew.cmu.edu/user/gl20/GeorgeLoewenstein/Papers_files/ pdf/PsychofCuriosity.pdf.

91   **"shields the searcher from such radical encounters":** Siva Vaidhyanathan, *The Googlization of Everything* (Berkeley and Los Angeles: University of California Press, 2011), 182.

91   "only give you answers": Pablo Picasso, as quoted in Gerd Leonhard, Media Futurist Web site, Dec. 8, 2004, accessed Feb. 9, 2011, www.mediafuturist.com/about.html.

92   "On Adderall, I was able to work": Joshua Foer, "The Adderall Me: My Romance with ADHD Meds," *Slate*, May 10, 2005, www.slate.com/id/2118315.

92   "pressures [to use enhancing drugs] are only going to grow": Margaret Talbot, "Brain Gain: The Underground World of 'Neuroenhancing Drugs,'" *New Yorker*, Apr. 27, 2009, accessed Dec. 14, 2010, www.newyorker.com/reporting/2009/04/27/090427fa_fact_talbot?currentPage=all.

93   "I think 'inside the box'": Erowid Experience Vaults, accessed Dec. 14, 2010, www.erowid.org/experiences/exp.php?ID=56716.

93   "a generation of very focused accountants": Talbot, "Brain Gain."

94   "an analogy no one has ever seen": Arthur Koestler, *Art of Creation* (New York: Arkana, 1989), 82.

94   "uncovers, selects, re-shuffles, combines, synthesizes": Ibid., 86.

95   the key to creative thought: Hans Eysenck, *Genius: The Natural History of Creativity* (Cambridge: Cambridge University Press, 1995).

95   box represents the solution horizon: Hans Eysenck, "Creativity and Personality: Suggestions for a Theory," *Psychological Inquiry*, 4, no. 3 (1993): 147–78.

97   no idea what they're looking for: Aharon Kantorovich and Yuval Ne'eman, "Serendipity as a Source of Evolutionary Progress in Science," *Studies in History and Philosophy of Science, Part A*, 20, no. 4: 505–29.

98   attach the candle to the wall: Karl Duncker, "On Problem Solving," *Psychological Monographs*, 58 (1945).

98   reluctance to "break perceptual set": George Katona, *Organizing and Memorizing* (New York: Columbia University Press, 1940).

99   creative people tend to see things: Arthur Cropley, *Creativity in Education and Learning* (New York: Longmans, 1967).

99   "sorted a total of 40 objects": N. J. C. Andreases and Pauline S. Powers, "Overinclusive Thinking in Mania and Schizophrenia," *British Journal of Psychology* 125 (1974): 452–56.

99   a "thing with weight": Cropley, *Creativity*, 39.

100   "Stop counting—there are 43 pictures": Richard Wiseman, *The Luck Factor* (New York: Hyperion, 2003), 43–44.

101   bilinguists are more creative than monolinguists: Charlan Nemeth and Julianne Kwan, "Minority Influence, Divergent Thinking and Detection of Correct Solutions," *Journal of Applied Social Psychology*, 17, I. 9 (1987): 1, accessed Feb. 7, 2011, http://onlinelibrary.wiley .com/doi/10.1111/j.1559-1816.1987.tb00339.x/abstract.

101   foreign ideas help us: W. M. Maddux, A. K. Leung, C. Chiu, and A. Galinsky, "Toward a More Complete Understanding of the Link Between Multicultural Experience and Creativity," *American Psychologist* 64 (2009): 156–58.

102   illustrates how creativity arises: Steven Johnson, *Where Good Ideas Come From: The Natural History of Innovation* (New York: Penguin, 2010), *ePub Bud*, accessed Feb 7, 2011, www.epubbud.com/read.php ?g=LN9DVC8S.

102   "wide and diverse sample of spare parts": Ibid., 6.

102   "environments that are powerfully suited": Ibid., 3.

102   "'serendipity' article in Wikipedia": Ibid., 13.

103   "shift from exploration and discovery": John Battelle, *The Search: How Google and Its Rivals Rewrote the Rules of Business and Transformed Our Culture* (New York: Penguin, 2005), 61.

103   "database of intentions": Ibid.

104   "We need help overcoming rationality": David Gelernter, *Time to Start Taking the Internet Seriously*, accessed Dec. 14, 2010, www.edge .org/3rd_culture/gelernter10/gelernter10_index.html.

105   "a vast island called California": Garci Rodriguez de Montalvo, *The Exploits of Esplandian* (Madrid: Editorial Castalia, 2003).

Chapter Four: The You Loop

109   "what a personal computer really is": Sharon Gaudin, "Total Recall: Storing Every Life Memory in a Surrogate Brain," *ComputerWorld*, Aug. 2, 2008, accessed Dec. 15, 2010, www.computerworld.com/s/ article/9074439/Total_Recall_Storing_every_life_memory_in_a _surrogate_brain.

109 **"You have one identity":** David Kirkpatrick, *The Facebook Effect: The Inside Story of the Company That Is Connecting the World* (New York: Simon and Schuster, 2010), 199.

109 **"I behave a different way":** "Live-Blog: Zuckerberg and David Kirkpatrick on the Facebook Effect," transcript of interview, *Social Beat*, accessed Dec. 15, 2010, http://venturebeat.com/2010/07/21/live-blog-zuckerberg-and-david-kirkpatrick-on-the-facebook-effect.

110 **"Same awkward self":** Ibid.

110 **that would be the norm:** Marshall Kirkpatrick, "Facebook Exec: All Media Will Be Personalized in 3 to 5 Years," *ReadWriteWeb*, Sept. 29, 2010, accessed Dec. 15, 2010, www.readwriteweb.com/archives/facebook_exec_all_media_will_be_personalized_in_3.php.

110 **"a world that all may enter":** John Perry Barlow, A Declaration of the Independence of Cyberspace, Feb. 8, 1996, accessed Dec. 15, 2010, https://projects.eff.org/~barlow/Declaration-Final.html.

111 **pseudonym with the real name:** Julia Angwin and Steve Stecklow, "'Scrapers' Dig Deep for Data on Web," *Wall Street Journal*, Oct. 12, 2010, accessed Dec. 15, 2010, http://online.wsj.com/article/SB10001424052748703358504575544381288117888.html.

111 **tied to the individual people who use them:** Julia Angwin and Jennifer Valentino-Devries, "Race Is On to 'Fingerprint' Phones, PCs," *Wall Street Journal*, Nov. 30, 2010, accessed Jan. 30, 2011, http://online.wsj.com/article/SB10001424052748704679204575646704100959546.html?mod=ITP_pageone_0.

112 **information sources make us freer:** Yochai Benkler, "Of Sirens and Amish Children: Autonomy, Information, and Law," *New York University Law Review*, 76 no. 23 (April 2001): 110.

115 **"more than the bits of data":** Daniel Solove, *The Digital Person: Technology and Privacy in the Information Age* (New York: New York University Press, 2004), 45.

116 **how someone behaves from who she is:** E. E. Jones and V. A. Harris, "The Attribution of Attitudes," *Journal of Experimental Social Psychology* 3 (1967): 1–24.

116 **electrocute other subjects:** Stanley Milgram, "Behavioral Study of Obedience," *Journal of Abnormal and Social Psychology* 67 (1963): 371–78.

116 **The plasticity of the self:** Paul Bloom, "First Person Plural," *Atlantic* (Nov. 2008), accessed Dec. 15, 2010, www.theatlantic.com/magazine/ archive/2008/11/first-person-plural/7055.

117 **aspirations played against their current desires:** Katherine L. Milkman, Todd Rogers, and Max H. Bazerman, "Highbrow Films Gather Dust: Time-Inconsistent Preferences and Online DVD Rentals," *Management Science* 55, no. 6 (June 2009): 1047–59, accessed Jan. 29, 2011, http:// opimweb.wharton.upenn.edu/documents/research/Highbrow.pdf.

117 **"want" movies like *Sleepless in Seattle*:** Milkman, et al., "Highbrow Films Gather Dust."

118 **"nuances of what it means to be human":** John Battelle, phone interview with author, Oct. 12, 2010.

118 **Google is working on it:** Jonathan McPhie, phone interview with author, Oct. 13, 2010.

119 **the "toxic knowledge" that might result:** Mark Rothstein, as quoted in Cynthia L. Hackerott, J.D., and Martha Pedrick, J.D., "Genetic Information Nondiscrimination Act Is a First Step; Won't Solve the Problem," Oct. 1, 2007, accessed Feb. 9, www.metrocorpcounsel.com/ current.php?artType=view&artMonth=January&artYear=2011&Entr yNo=7293.

119 **"The digital ghost of Jay Gatz":** Siva Vaidyanathan, "Naked in the 'Nonopticon,'" *Chronicle Review* 54, no. 23: B7.

120 **"high cognition" arguments:** Dean Eckles, phone interview with author, Nov. 9, 2010.

120 **increase the effectiveness of marketing:** Ibid.

122 **pitches framed as sweepstakes:** PK List Marketing, "Free to Me— Impulse Buyers," accessed Jan. 28, 2011, www.pklistmarketing.com/ Data%20Cards/Opportunity%20Seekers%20&%20Sweepstakes%20 Participants/Cards/Free%20To%20Me%20-%20Impulse%20Buyers .htm.

123   **"smartphone to be doing searches constantly"**: Robert Andrews, "Google's Schmidt: Autonomous, Fast Search Is 'Our New Definition,'" *paidContent*, Sept. 7, 2010, accessed Dec. 15, 2010, http://paidcontent.co.uk/article/419-googles-schmidt-autonomous-fast-search-is-our-new-definition.

124   **"'Not-So-Minimal' Consequences of Television News"**: Shanto Iyengar, Mark D. Peters, and Donald R. Kinder, "Experimental Demonstrations of the 'Not-So-Minimal' Consequences of Television News Programs," *American Political Science Review* 76, no. 4 (1982): 848–58.

124   **"believe that defense or pollution"**: Ibid.

124   **strength of this priming effect**: Drew Westen, *The Political Brain: The Role of Emotion in Deciding the Fate of the Nation* (Cambridge, MA: Perseus, 2007).

125   **study by Hasher and Goldstein**: Lynn Hasher and David Goldstein, "Frequency and the Conference of Referential Validity," *Journal of Verbal Learning and Verbal Behaviour* 16 (1977): 107–12.

126   **"surrounded by downward-sloping land"**: Matt Cohler, phone interview with author, Nov. 23, 2010.

128   **results had been randomly redistributed**: Robert Rosenthal and Lenore Jacobson, "Teachers' Expectancies: Determinants of Pupils' IQ Gains," *Psychological Reports*, 19 (1966): 115–18.

129   **"network-based categorizations"**: Dalton Conley, *Elsewhere, U.S.A.: How We Got from the Company Man, Family Dinners, and the Affluent Society to the Home Office, BlackBerry Moms, and Economic Anxiety* (New York: Pantheon, 2008), 164.

130   **"Model-T version of what's possible"**: Geoff Duncan, "Netflix Offers $1Mln for Good Movie Picks," *Digital Trends*, Oct. 2, 2006, accessed Dec. 15, 2010, www.digitaltrends.com/computing/netflix-offers-1-mln-for-good-movie-picks.

130   **"a PC and some great insight"**: Katie Hafner, "And If You Liked the Movie, a Netflix Contest May Reward You Handsomely," *New York Times*, Oct. 2, 2006, accessed Dec. 15, 2010, www.nytimes.com/2006/10/02/technology/02netflix.html.

131 **success using social-graph data:** Charlie Stryler, Marketing Panel at 2010 Social Graph Symposium, Microsoft Campus, Mountain View, CA, May 21, 2010.

132 **"the creditworthiness of your friends":** Julia Angwin, "Web's New Gold Mine," *Wall Street Journal*, July 30, 2010, accessed on Feb. 7, 2011, http://online.wsj.com/article/SB10001424052748703940904575395073512989404.html.

133 **reality doesn't work that way:** David Hume, *An Enquiry Concerning Human Understanding*, Harvard Classics, volume 37, Section VII, Part I, online edition, (P. F. Collier & Son; 1910), accessed Feb. 7, 2011, http://18th.eserver.org/hume-enquiry.html.

133 **purpose of science, for Popper:** Karl Popper, *The Logic of Scientific Discovery* (New York: Routledge, 1992).

135 **"no more incidents or adventures in the world":** Fyodor Dostoevsky, *Notes from Underground*, trans. Richard Pevear and Laura Volokhonsky (New York: Random House, 1994), 24.

Chapter Five: The Public Is Irrelevant

137 **"others who see what we see":** Hannah Arendt, *The Portable Hannah Arendt* (New York: Penguin, 2000), 199.

137 **"neutralize the influence of the newspapers":** Alexis de Tocqueville, *Democracy in America* (New York: Penguin, 2001).

138 **"a gross violation of Chinese sovereignty":** "NATO Hits Chinese Embassy," *BBC News*, May 8, 1999, accessed Dec. 17, 2010, http://news.bbc.co.uk/2/hi/europe/338424.stm.

138 **"most vital are the largely anonymous online forums":** Tom Downey, "China's Cyberposse," *New York Times*, Mar. 3, 2010, accessed Dec. 17, 2010, www.nytimes.com/2010/03/07/magazine/07Human-t.html?pagewanted=1.

138 **"an elite, wired section of the population":** Shanthi Kalathil and Taylor Boas, "Open Networks, Closed Regimes: The Impact of the Internet on Authoritarian Rule," *First Monday* 8, no. 1–6 (2003).

139 **"Shareholders want to make money":** Clive Thompson, "Google's China

Problem (and China's Google Problem)," *New York Times*, Apr. 23, 2006, accessed Dec. 17, 2010, www.nytimes.com/2006/04/23/magazine/23google.html.

139 **"What the government cares about":** James Fallows, "The Connection Has Been Reset," *Atlantic*, Mar. 2008, accessed Dec. 17, 2010, www.theatlantic.com/magazine/archive/2008/03/-ldquo-the-connection-has-been-reset-rdquo/6650.

139 **"peer pressure, and self-censorship":** Fallows, "Connection Has Been Reset."

140 **"sense that they're looking at everything":** Thompson, "Google's China Problem."

140 **"Internet Police will maintain order":** Hong Yan, "Image of Internet Police: JingJang and Chacha Online," *China Digital Times*, Feb. 8, 2006, accessed Dec. 17, 2010, http://chinadigitaltimes.net/china/internet-police/page/2.

140 **"see my friends, live happily":** Thompson, "Google's China Problem."

140 **"if Internet users have some porn":** Associated Press, "Web Porn Seeps Through China's Great Firewall," July 22, 2010, accessed Dec. 17, 2010, www.cbsnews.com/stories/2010/07/22/tech/main6703860.shtml.

141 **"trying to nail Jell-O to the wall":** Bill Clinton, "America's Stake in China," *Blueprint*, June 1, 2000, accessed Dec. 17, 2010, www.dlc.org/ndol_ci.cfm?kaid=108&subid=128&contentid=963.

142 **"able to get handheld American flags?":** Laura Miller and Sheldon Rampton, "The Pentagon's Information Warrior: Rendon to the Rescue," *PR Watch* 8, no. 4 (2001).

142 **"border patrols [are] replaced by beaming patrols":** John Rendon, as quoted in Franklin Foer, "Flacks Americana," *New Republic*, May 20, 2002, accessed Feb. 9, 2011, www.tnr.com/article/politics/flacks-americana?page=0,2.

142 **thesaurus:** John Rendon, phone interview by author, Nov. 1, 2010.

143 **"consume, distribute, and create":** Eric Schmidt and Jared Cohen, "The Digital Disruption: Connectivity and the Diffusion of Power," *Foreign Affairs* (Nov.–Dec. 2010).

144 **Flatow was an Olympic gymnast:** Stephen P. Halbrook, "'Arms in the Hands of Jews Are a Danger to Public Safety': Nazism, Firearm Registration, and the Night of the Broken Glass, *St. Thomas Law Review* 21 (2009): 109–41, 110, www.stephenhalbrook.com/law_review_articles/ Halbrook_macro_final_3_29.pdf.

145 **the cloud "is actually just a handful of companies":** Clive Thompson, interview with author, Brooklyn, NY, Aug. 13, 2010.

145 **there was nowhere to go:** Peter Svensson, "WikiLeaks Down? Cables Go Offline After Site Switches Servers," *Huffington Post*, Dec. 1, 2010, accessed Feb. 9, 2011, www.huffingtonpost.com/2010/12/01/wikile aks-down-cables-go-_n_790589.html.

145 **"lose your constitutional protections immediately":** Christopher Ketcham and Travis Kelly, "The More You Use Google, the More Google Knows About You," *AlterNet*, Apr. 9, 2010, accessed Dec. 17, 2010, www.alternet.org/investigations/146398/total_information_aware ness:_the_more_you_use_google,_the_more_google_knows_about_you _?page=entire.

146 **"cops will love this":** "Does Cloud Computing Mean More Risks to Privacy?," *New York Times*, Feb. 23, 2009, accessed Feb. 8, 2011, http://bits .blogs.nytimes.com/2009/02/23/does-cloud-computing-mean-more -risks-to-privacy.

146 **the three companies quickly complied:** Antone Gonsalves, "Yahoo, MSN, AOL Gave Search Data to Bush Administration Lawyers," *Information Week*, Jan. 19, 2006, accessed Feb. 9, 2011, www.information week.com/news/security/government/showArticle.jhtml?articleID =177102061.

146 **predict future real-world events:** Ketcham and Kelly, "The More You Use Google."

146 **"an individual must increasingly give information":** Jonathan Zittrain, *The Future of the Internet—and How to Stop It* (New Haven: Yale University Press, 2008), 201.

147 **"an implicit bargain in our behavior":** John Battelle, phone interview with author, Oct. 12, 2010.

147 **"redistribution of information power"**: Viktor Mayer-Schonberger, *Delete: The Virtue of Forgetting in the Digital Age* (Princeton: Princeton University Press, 2009), 107.

148 **real-world violence:** George Gerbner, "TV Is Too Violent Even Without Executions," *USA Today*, June 16, 1994, 12A, accessed Feb. 9, 2011 through LexisNexis.

149 **"who tells the stories of a culture":** "Fighting 'Mean World Syndrome,'" *GeekMom* blog, *Wired*, Jan. 27, 2011, accessed Feb. 9, 2011, www.wired .com/geekdad/2011/01/fighting-%E2%80%9Cmean-world-syn drome%E2%80%9D/.

149 **friendly world syndrome:** Dean Eckles, "The 'Friendly World Syndrome' Induced by Simple Filtering Rules," *Ready-to-Hand: Dean Eckles on People, Technology, and Inference* blog, Nov. 10, 2010, accessed Feb. 9, 2011, www.deaneckles.com/blog/386_the-friendly-world-syn drome-induced-by-simple-filtering-rules/.

149 **gravitated toward Like:** "What's the History of the Awesome Button (That Eventually Became the Like Button) on Facebook?" Quora Forum, accessed Dec. 17, 2010, www.quora.com/Facebook-company/ Whats-the-history-of-the-Awesome-Button-that-eventually-became -the-Like-button-on-Facebook.

151 **"against the cruise line industry":** Hollis Thomases, "Google Drops Anti-Cruise Line Ads from AdWords," Web Ad.vantage, Feb. 13, 2004, accessed Dec. 17, 2010, www.webadvantage.net/webadblog/ google-drops-anti-cruise-line-ads-from-adwords-338.

151–52 **identify who was persuadable:** "How Rove Targeted the Republi-can Vote," *Frontline*, accessed Feb. 8, 2011, www.pbs.org/wgbh/pages/ frontline/shows/architect/rove/metrics.html.

152 **"Amazon's recommendation engine is the direction":** Mark Steitz and Laura Quinn, "An Introduction to Microtargeting in Politics," accessed Dec. 17, 2010, www.docstoc.com/docs/43575201/An-Introduction-to -Microtargeting-in-Politics.

153 **round-the-clock "war room":** "Google's War Room for the Home Stretch of Campaign 2010," e.politics, Sept. 24, 2010, accessed Feb. 9,

2011, www.epolitics.com/2010/09/24/googles-war-room-for-the-home
-stretch-of-campaign-2010/.

155 **"campaign wanted to spend on Facebook":** Vincent R. Harris, "Facebook's Advertising Fluke," *TechRepublican*, Dec. 21, 2010, accessed Feb. 9, 2011, http://techrepublican.com/free-tagging/vincent-harris.

155 **have the ads pulled off the air:** Monica Scott, "Three TV Stations Pull 'Demonstrably False' Ad Attacking Pete Hoekstra," *Grand Rapids Press*, May 28, 2010, accessed Dec. 17, 2010, www.mlive.com/politics/index.ssf/2010/05/three_tv_stations_pull_demonst.html.

157 **"improve the likelihood that a registered Republican":** Bill Bishop, *The Big Sort: Why the Clustering of Like-Minded America Is Tearing Us Apart* (New York: Houghton Mifflin, 2008), 195.

157 **"likely to be most salient in the politics":** Ronald Inglehart, *Modernization and Postmodernization* (Princeton: Princeton University Press, 1997), 10.

159 **Pabst began to sponsor hipster events:** Neal Stewart, "Marketing with a Whisper," *Fast Company*, Jan. 11, 2003, accessed Jan. 30, 2011, www.fastcompany.com/fast50_04/winners/stewart.html.

159 **"$44 in US currency":** Max Read, "Pabst Blue Ribbon Will Run You $44 a Bottle in China," *Gawker*, July 21, 2010, accessed Feb. 9, 2011, http://m.gawker.com/5592399/pabst-blue-ribbon-will-run-you-44-a-bottle-in-china.

160 **"I serve as a blank screen":** Barack Obama, *The Audacity of Hope: Thoughts on Reclaiming the American Dream* (New York: Crown, 2006), 11.

161 **"We lose all perspective":** Ted Nordhaus, phone interview with author, Aug. 31, 2010.

162 **"the source is basically in thought":** David Bohm, *Thought as a System* (New York: Routledge, 1994) 2.

163 **"participants in a pool of common meaning":** David Bohm, *On Dialogue* (New York: Routledge, 1996), x–xi.

164 **"define and express its interests":** John Dewey, *The Public and Its Problems* (Athens, OH: Swallow Press, 1927), 146.

Chapter Six: Hello, World!

165 "no intelligence or skill in navigation": Plato, *First Alcibiades*, in *The Dialogues of Plato*, vol. 4, trans. Benjamin Jowett (Oxford, UK: Clarendon Press, 1871), 559.

166 "We are as Gods": Stewart Brand, *Whole Earth Catalog* (self-published, 1968), accessed Dec. 16, 2010, http://wholeearth.com/issue/1010/article/195/we.are.as.gods.

167 "make any man (or woman) a god": Steven Levy, *Hackers: Heroes of the Computer Revolution* (New York: Penguin, 2001), 451.

167 "having some troubles with my family": "How Eliza Works," accessed Dec. 16, 2010, http://chayden.net/eliza/instructions.txt.

168 "way of acting without consequence": Siva Vaidyanathan, phone interview with author, Aug. 9, 2010.

168 "not a very good program": Douglas Rushkoff, interview with author, New York, NY, Aug. 25, 2010.

168 "politics tends to be seen by programmers": Gabriella Coleman, "The Political Agnosticism of Free and Open Source Software and the Inadvertent Politics of Contrast," *Anthropological Quarterly*, 77, no. 3 (Summer 2004): 507–19, Academic Search Premier, EBSCOhost.

170 "addictive control as well": Levy, *Hackers*, 73.

172 "Howdy" is a better opener than "Hi": Christian Rudder, "Exactly What to Say in a First Message," Sept. 14, 2009, accessed Dec. 16, 2010, http://blog.okcupid.com/index.php/online-dating-advice-exactly-what-to-say-in-a-first-message.

173 "hackers don't tend to know any of that": Steven Levy, "The Unabomber and David Gelernter," *New York Times*, May 21, 1995, accessed Dec. 16, 2010, www.unabombers.com/News/95-11-21-NYT.htm.

174 "engineering relationships among people": Langdon Winner, "Do Artifacts Have Politics?" *Daedalus* 109, no. 1 (Winter 1980): 121–36.

175 "code is law": Lawrence Lessig, *Code*. (New York: Basic Books, 2006).

175 **"choose structures for technologies":** Winner, "Do Artifacts Have Politics."

176 **Hacker Jargon File:** The Jargon File, Version 4.4.7, Appendix B. A Portrait of J. Random Hacker, accessed Feb. 9, 2011, http://linux .rz.ruhr-uni-bochum.de/jargon/html/politics.html.

177 **"social utility" as if it's a twenty-first-century phone company:** Mark Zuckerberg executive bio, Facebook press room, accessed on Feb. 8, 2011, http://www.facebook.com/press/info.php?execbios.

178 **"come to Google because they choose to":** Greg Jarboe, "A 'Fireside Chat' with Google's Sergey Brin," Search Engine Watch, Oct. 16, 2003, accessed Dec. 16, 2010, http://searchenginewatch.com/3081081.

178 **"the future will be personalized":** Gord Hotckiss, "Just Behave: Google's Marissa Mayer on Personalized Search," Searchengineland, Feb. 23, 2007, accessed Dec. 16, 2010, http://searchengineland.com/ just-behave-googles-marissa-mayer-on-personalized-search-10592.

179 **"It's technology, not business or government":** David Kirpatrick, "With a Little Help from his Friends," *Vanity Fair* (Oct. 2010), accessed Dec. 16, 2010, www.vanityfair.com/culture/features/2010/10/sean-parker -201010.

179 **"seventh kingdom of life":** Kevin Kelly, *What Technology Wants* (New York: Viking, 2010).

180 **"shirt or fleece that I own":** Mark Zuckerberg, remarks to Startup School Conference, *XConomy*, Oct. 18, 2010, accessed Feb. 8, 2010, www.xconomy.com/san-francisco/2010/10/18/mark-zuckerberg -goes-to-startup-school-video//.

181 **"'the rest of the world is wrong'":** David A. Wise and Mark Malseed, *The Google Story* (New York: Random House, 2005), 42.

182 **"tradeoffs with success in other domains":** Jeffrey M. O'Brien, "The Pay-Pal Mafia," *Fortune*, Nov. 14, 2007, accessed Dec. 16, 2010, http://money .cnn.com/2007/11/13/magazines/fortune/paypal_mafia.fortune/index2. htm.

183 **sold to eBay for $1.5 billion:** Troy Wolverton, "It's official: eBay Weds PayPal," *CNET News*, Oct. 3, 2002, accessed Dec. 16, 2010, http://news .cnet.com/Its-official-eBay-weds-PayPal/2100-1017_3-960658.html.

183   **"impact and force change"**: Peter Thie, "Education of a Libertarian,"
      Cato Unbound, Apr. 13, 2009, accessed Dec. 16, 2010, www.cato
      -unbound.org/2009/04/13/peter-thiel/the-education-of-a-libertarian.

183   **"end the inevitability of death and taxes"**: Chris Baker, "Live Free or
      Drown: Floating Utopias on the Cheap," *Wired*, Jan. 19, 2009, accessed
      Dec. 16, 2010, www.wired.com/techbiz/startups/magazine/17-02/mf
      _seasteading?currentPage=all.

183   **"'capitalist democracy' into an oxymoron"**: Thiel, "Education of a Lib-
      ertarian."

184   **"makes a living being against computers"**: Nicholas Carlson, "Peter
      Thiel Says Don't Piss Off the Robots (or Bet on a Recovery)," *Business
      Insider*, Nov. 18, 2009, accessed Dec. 16, 2010, www.businessinsider
      .com/peter-thiel-on-obama-ai-and-why-he-rents-his-mansion
      -2009-11#.

184   **"which technologies to foster"**: Ronald Bailey, "Technology Is at the
      Center," Reason.com, May 2008, accessed Dec. 16, 2010, http://
      reason.com/archives/2008/05/01/technology-is-at-the-center/
      singlepage.

184   **"way I think about the business"**: Deepak Gopinath, "PayPal's Thiel
      Scores 230 Percent Gain with Soros-Style Fund," CanadianHedgeWatch
      .com, Dec. 4, 2006, accessed Jan. 30, 2011, at www.canadianhedge
      watch.com/content/news/general/?id=1169.

184   **"that voting will make things better"**: Peter Thiel, "Your Suffrage
      Isn't in Danger. Your Other Rights Are," *Cato Unbound*, May 1, 2009,
      accessed Dec. 16, 2010, www.cato-unbound.org/2009/05/01/peter
      -thiel/your-suffrage-isnt-in-danger-your-other-rights-are.

185   **talked to Scott Heiferman**: Interview with author, New York, NY,
      Oct. 5, 2010.

188   **"good or bad, nor is it neutral"**: Melvin Kranzberg, "Technology and
      History: 'Kranzberg's Laws,'" *Technology and Culture* 27, no. 3 (1986):
      544–60.

Chapter Seven: What You Want, Whether You Want It or Not

189   "millions of people doing complicated things": Noah Wardrip-Fruin and Nick Montfort, *The New Media Reader*, Vol. 1 (Cambridge: MIT Press, 2003), 8.

189   "yet to be completely correlated": Isaac Asimov, *The Intelligent Man's Guide to Science* (New York: Basic Books, 1965),

190   "you've got a problem": Bill Jay, phone interview with author, Oct. 10, 2010.

191   ads tailored to her: Jason Mick, "Tokyo's 'Minority Report' Ad Boards Scan Viewer's Sex and Age," Daily Tech, July 16, 2010, accessed Dec. 17, 2010, www.dailytech.com/Tokyos+Minority+Report+Ad+Boards +Scan+Viewers+Sex+and+Age/article19063.htm.

191   the future of art: David Shields, *Reality Hunger: A Manifesto* (New York: Knopf, 2010). Credit to Michiko Kakutani, whose review led me to this book.

193   interrogated by a virtual agent: M. Ryan Calo, "People Can Be So Fake: A New Dimension to Privacy and Technology Scholarship," *Penn State Law Review* 114 , no. 3 (2010): 810–55.

193   Kismet increased donations by 30 percent: Vanessa Woods, "Pay Up, You Are Being Watched," *New Scientist*, Mar. 18, 2005, accessed Dec. 17, 2010, www.newscientist.com/article/dn7144-pay-up-you-are -being-watched.html.

193   "Computers programmed to be polite": Calo, "People Can Be So Fake."

194   "not evolved to twentieth-century technology": Ibid.

195   identity and criminal record in seconds: Maureen Boyle, "Video: Catching Criminals? Brockton Cops Have an App for That," *Brockton Patriot Ledger*, June 15, 2010, accessed Dec. 17, 2010, www.patriot ledger.com/news/cops_and_courts/x1602636300/Catching -criminals-Cops-have-an-app-for-that.

195   "other images of you with ninety-five percent accuracy": Jerome Taylor, "Google Chief: My Fears for Generation Facebook," *Independent*, Aug. 18, 2010, accessed Dec. 17, 2010, www.independent.co.uk/life-style/

gadgets-and-tech/news/google-chief-my-fears-for-generation-face
book-2055390.html.

197 "The future is already here": William Gibson, interview on NPR's
     Fresh Air, Aug. 31, 1993, accessed Dec. 17, 2010, www.npr.org/
     templates/story/story.php?storyId=1107153.

197 your identity already tagged: "RFID Bracelet Brings Facebook to the Real
     World," Aug. 20, 2010, accessed Dec. 17, 2010, www.psfk.com/2010/08/
     rfid-bracelet-brings-facebook-to-the-real-world.html.

198 "real world that can be indexed": Reihan Salam, "Why Amazon Will Win
     the Internet," *Forbes*, July 30, 2010, accessed Dec. 17, 2010, www.forbes
     .com/2010/07/30/amazon-kindle-economy-environment-opinions
     -columnists-reihan-salam.html.

198 "some have termed 'smart dust'": David Wright, Serge Gutwirth,
     Michael Friedewald, Yves Punie, and Elena Vildjiounaite, *Safeguards in
     a World of Ambient Intelligence* (Berlin/Dordrecht: Springer Science,
     2008): abstract.

199 four-year joint effort: Google/Harvard press release. "Digitized Book
     Project Unveils a Quantitative 'Cultural Genome,'" accessed Feb. 8,
     2011, http://www.seas.harvard.edu/news-events/news-archive/2010/
     digitized-books.

200 "censorship and propaganda": Ibid.

200 nearly sixty languages: Google Translate Help Page, accessed Feb. 8,
     2011, http://translate.google.com/support/?hl=en.

201 better and better: Nikki Tait, "Google to translate European patent
     claims," *Financial Times*, Nov. 29, 2010, accessed Feb. 9, 2010, www
     .ft.com/cms/s/0/02f71b76-fbce-11df-b79a-00144feab49a.html.

202 "what to do with them": Danny Sullivan, phone interview with author,
     Sept. 10, 2010.

202 "flash crash": Graham Bowley, "Stock Swing Still Baffles, with an Omi-
     nous Tone," *New York Times*, Aug. 22, 2010, accessed Feb. 8, 2010,
     www.nytimes.com/2010/08/23/business/23flash.html.

202 provocative article in *Wired*: Chris Anderson, "The End of Theory:
     The Data Deluge Makes the Scientific Method Obsolete," *Wired*, June

23, 2008, accessed Feb. 10, 2010, http://www.wired.com/science/
discoveries/magazine/16-07/pb_theory.

203 **greatest achievement of human technology:** Hillis quoted in Jennifer
Riskin, *Genesis Redux: Essays in the History and Philosophy of Artificial
Life* (Chicago: University of Chicago Press, 2007), 200.

204 **"advertiser-funded media":** Marisol LeBron, "'Migracorridos': Another
Failed Anti-immigration Campaign," North American Congress of
Latin America, Mar. 17, 2009, accessed Dec. 17, 2010, https://nacla
.org/node/5625.

205 **characters using the companies' products throughout:** Mary McNa-
mara, "Television Review: 'The Jensen Project,'" *Los Angeles Times*,
July 16, 2010, accessed Dec. 17, 2010, http://articles.latimes.
com/2010/jul/16/entertainment/la-et-jensen-project-20100716.

205 **product-placement hooks throughout:** Jenni Miller, "Hansel and Gretel in 3D? Yeah, Maybe." *Moviefone* blog, July 19, 2010, accessed Dec.
17, 2010, http://blog.moviefone.com/2010/07/19/hansel-and-gretel-in
-3d-yeah-maybe.

205 **the corporate owner of Lipslicks:** Motoko Rich, "Product Placement
Deals Make Leap from Film to Books," *New York Times*, Nov. 9, 2008,
accessed Dec. 17, 2010, www.nytimes.com/2008/02/19/arts/19iht-20
bookplacement.10177632.html?pagewanted=all.

207 **increase "purchase intentions" by 21 percent:** John Hauser and Glen
Urban, "When to Morph," Aug. 2010, accessed Dec. 17, 2010, http://
web.mit.edu/hauser/www/Papers/Hauser-Urban-Liberali_When_to_
Morph_Aug_2010.pdf.

207 **"turn it into useful information":** Jane Wardell, "Raytheon Unveils Scorpion Helmet Technology," Associated Press, July 23, 2010, accessed
Dec. 17, 2010 at www.boston.com/business/articles/2010/07/23/
raytheon_unveils_scorpion_helmet_technology.

208 **"turns the whole world into a display":** Wardell, "Raytheon Unveils
Scorpion Helmet Technology."

208 **TV experience overlaid on a real game:** Michael Schmidt, "To Pack a
Stadium, Provide Video Better Than TV," *New York Times*, July 28,

2010, accessed Dec. 17, 2010, www.nytimes.com/2010/07/29/sports/football/29stadium.html?_r=1.

208 **AugCog, which uses cognitive neuroscience:** Augmented Cognition International Society Web site, accessed Dec. 17, 2010, www.augmented cognition.org.

209 **500 percent increase in working memory:** "Computers That Read Your Mind," *Economist*, Sept. 21, 2006, accessed Dec. 17, 2010, www.economist.com/node/7904258?story_id=7904258.

209 **at least sixteen different ways:** Gary Hayes, "16 Top Augmented Reality Business Models," *Personalize Media* (Gary Hayes's blog), Sept. 14, 2009, accessed Dec. 17, 2010, www.personalizemedia.com/16-top-augmented-reality-business-models.

210 **solve problems for people:** Chris Coyne, interview with author, New York, NY, Oct. 6, 2010.

211 **"reality" is "one of the few words":** Vladimir Nabokov, *Lolita* (New York: Random House, 1997), 312.

213 **powering the marketing campaigns:** David Wright et al., *Safeguards in a World of Ambient Intelligence* (London: Springer, 2008), 66, accessed through Google eBooks, Feb. 8, 2011.

214 **"machines make more of their decisions":** Bill Joy, "Why the Future Doesn't Need Us," *Wired* (Apr. 2000) accessed Dec. 17, 2010, www.wired.com/wired/archive/8.04/joy.html.

Chapter Eight: Escape from the City of Ghettos

217 **"the nature of his own person":** Christopher Alexander et al., *A Pattern Language* (New York: Oxford University Press, 1977), 8.

217 **"Long Live the Web"** Sir Tim Berners-Lee, "Long Live the Web: A Call for Continued Open Standards and Neutrality," *Scientific American*, Nov. 22, 2010.

219 **"need to address the core issues":** Bill Joy, phone interview with author, Oct. 1 2010.

220 **ideal nook for kids:** Alexander et al., *A Pattern Language*, 445, 928–29.

220 **"distinct pattern language":** Ibid., xvi.

220 "city of ghettos": Ibid., 41–43.

221 "dampens all significant variety": Ibid., 43.

221 "move easily from one to another": Ibid., 48.

221 "support for his idiosyncrasies": Ibid.

222 "psychological equivalent of obesity": danah boyd. "Streams of Content, Limited Attention: The Flow of Information through Social Media," *Web2.0 Expo.* New York, NY: Nov. 17, 2007, accessed July 19, 2008, www.danah.org/papers/talks/Web2Expo.html.

223 how to build a better mousetrap: "A Better Mousetrap," *This American Life* no. 366, aired Oct. 10, 2008, www.thisamericanlife.org/radio -archives/episode/366/a-better-mousetrap-2008.

223 you'll catch your mouse: Ibid.

223 "jumping out of that recursion loop": Matt Cohler, phone interview with author, Nov. 23, 2010.

226 organ donation rates in different European countries: Dan Ariely as quoted in Lisa Wade, "Decision Making and the Options We're Offered," *Sociological Images blog,* Feb. 17, 2010, accessed Dec. 17, 2010, http:// thesocietypages.org/socimages/2010/02/17/decision-making-and-the -options-were-offered/.

229 "only when regulation is transparent": Lawrence Lessig, *Code* (New York: Basic Books, 2006), 260, http://books.google.com/books?id=lm XIMZiU8yQC&pg=PA260&lpg=PA260&dq=lessig+political+response +transparent+code&source=bl&ots=wR0WRuJ61u&sig=iSIiM0pnEaf -o5VPvtGcgXXEeL8&hl=en&ei=1bI0TfykGsH38Ab7-tDJCA&sa =X&oi=book_result&ct=result&resnum=1&ved=0CBcQ6AEwAA#v =onepage&q&f=false.

230 "one of the world's worst kept secrets": Amit Singhal, "Is Google a Monopolist? A Debate," Opinion Journal, *Wall Street Journal,* Sept. 17, 2010, http://online.wsj.com/article/SB1000142405274870346670450 75489582364177978.html?mod=googlenews_wsj#U301271935 9440EB.

231 "honest and objective about ourselves": "Philip Foisie's memos to the management of the *Washington Post,*" Nov. 10, 1969, accessed Dec. 20, 2010, http://newsombudsmen.org/articles/origins/article-1-mcgee.

231   **"the common good":** Arthur Nauman, "News Ombudsmanship: Its
      Theory and Rationale," Press Regulation: How Far Has it Come?
      symposium, Seoul, South Korea, June 1994.

232   **that this expectation is one that ... most Americans share:** Jeffrey
      Rosen, "The Web Means the End of Forgetting," *New York Times Maga-*
      zine, July 21, 2010, www.nytimes.com/2010/07/25/magazine/25privacy
      -t2.html?_r=1&pagewanted=all.

235   **"help it find a larger audience":** Author interview with confidential
      source.

237   **Google is just a company:** "Transcript: Stephen Colbert Interviews
      Google's Eric Schmidt on *The Colbert Report,*" Search Engine Land,
      Sept. 22, 2010, accessed Dec. 20, 2010, http://searchengineland.com/
      googles-schmidt-colbert-report-51433.

237   **expose their audiences to both sides:** Cass R. Sunstein, *Republic*
      .com (Princeton: Princeton University Press, 2001).

240   **"we shouldn't have to accept":** Caitlin Petre phone interview with
      Marc Rotenberg, Nov. 5, 2010.

241   **and 70 percent do:** "Mistakes Do Happen: Credit Report Errors Mean
      Consumers Lose," US PIRG, accessed Feb. 8, 2010, http://www.uspirg
      .org/home/reports/report-archives/financial-privacy–security/financial
      -privacy-security/mistakes-do-happen-credit-report-errors-mean
      -consumers-lose.

INDEX